FEED SACKS

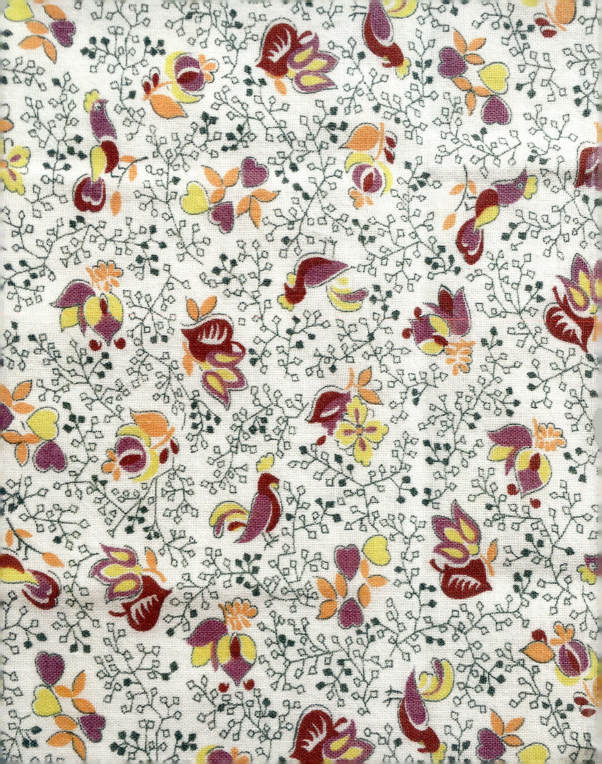

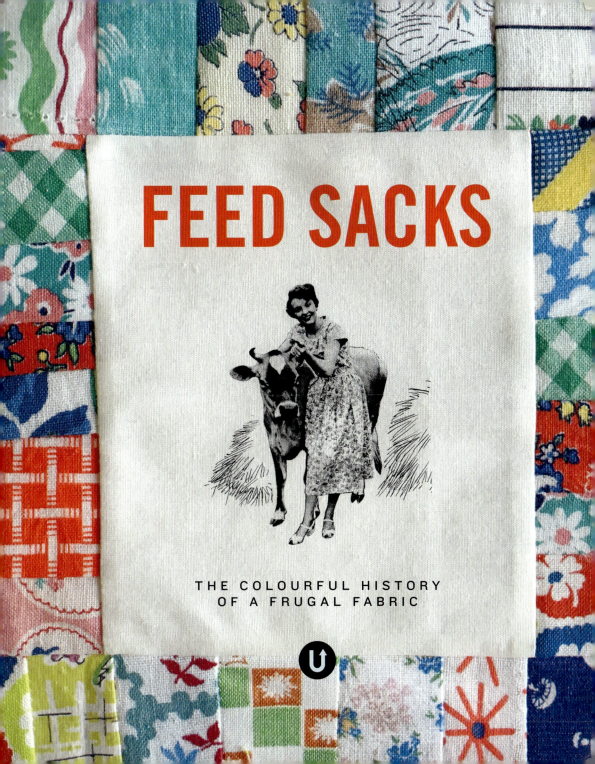

FEED SACKS

THE COLOURFUL HISTORY
OF A FRUGAL FABRIC

UPPERCASE PUBLISHING INC

201b – 908, 17th Ave SW, Calgary, Alberta, Canada T2T 0A3

uppercasemagazine.com

WRITER **Linzee Kull McCray**

EDITOR / IMAGE RESEARCH / BOOK DESIGN **Janine Vangool**

COPYEDITING **Correy Baldwin**

Full sacks and pattern swatches are courtesy the following collections: pp 6–7 collection of the author; pp 8–10 Sharon Forth; p11 Gloria Hall; p 16 Janine Vangool; p19 Sharon Forth; p30 Janine Vangool; pp 63–65 Gloria Hall; pp 68–69 Paul Pugsley; p70 Paul Pugsley except for Quaker Pig-N-Hog, Linzee Kull McCray and Ful-O-Pep, Pam Ehrhardt; pp 74–75 Janine Vangool, p77 Linzee Kull McCray; p78 Paul Pugsley; p79 Janine Vangool; pp 80–90 Paul Pugsley; p91 Paul Pugsley except for Great Western Pure Sugar, collection of the author; pp 94–95 Linzee Kull McCray; p101 Paul Pugsley; pp 105, 111, 114–115, 124–125 Gloria Hall; p 127 Paul Pugsley; pp 128–129 Amy Barickman; p129 baby's bib, Gloria Hall; pp 130–135 Gloria Hall; p 136 Paul Pugsley, p137 Gloria Hall; p 138 Amy Barickman; p 140–141 Janine Vangool; pp 142–151 Gloria Hall; pp 153–155 Gloria Hall; p159 Janine Vangool, p160 Sharon Forth; pp 166–167 Janine Vangool; p169 Gloria Hall; pp 176–179 Paul Pugsley; pp 180–181 Gloria Hall; p182 Paul Pugsley; p184 Janine Vangool; p186 Sharon Forth; p188 Gloria Hall; p196 Sharon Forth; p198 Gloria Hall; pp 200–203 Janine Vangool; p204 Paul Pugsley; p205 Gloria Hall; p206 Paul Pugsley; p207 Gloria Hall; pp 208–211 Paul Pugsley; pp214–215 Gloria Hall; p218 Sharon Forth; p219 Janine Vangool; pp 222, 225, 226 Sharon Forth; p227 Paul Pugsley, pp 229–230 Linzee Kull McCray; pp 232–233 Janine Vangool, p235 Gloria Hall; p236 Sharon Forth; pp 239–240 Janine Vangool; pp 242–245 Gloria Hall; p251 Janine Vangool except for framed certificate, Gloria Hall; pp 253–257 Paul Pugsley; pp 258, 262, 264, 268–269 Janine Vangool; pp 270–271 Gloria Hall; p282 Linzee Kull McCray; pp 283–284 Gloria Hall; p285 Janine Vangool; pp 286–291 Gloria Hall; p293 Janine Vangool; p296 Linzee Kull McCray; pp 297–298 Janine Vangool; p300 Gloria Hall; p301 Janine Vangool; p302 Paul Pugsley; pp 303–304 Gloria Hall; p305 Paul Pugsley; pp 306–307 Gloria Hall; pp 310–311 Linzee Kull McCray; p314 Paul Pugsley; p315 Gloria Hall except Kimbell sack, Amy Barickman; pp 316–317 Paul Pugsley; pp 318–332 Janine Vangool; p334 Janine Vangool; p335 Gloria Hall; p338 Janine Vangool; p339 Gloria Hall; pp 340–344 Janine Vangool; pp 345–349 Gloria Hall; pp 350–351 Janine Vangool; pp 354–355 Denyse Schmidt; pp 356–357 Amy Barickman; p359 (left) Janine Vangool; p359 (right) Linzee Kull McCray; p360 Linzee Kull McCray; pp 361–363 Janine Vangool; p365 Sharon Forth; pp 366–367 Janine Vangool; pp 368–369 Gloria Hall except lower right, Paul Pugsley; pp 370–391 Gloria Hall; pp 392–413 Paul Pugsley; pp 414–425 International Quilt Study Center and Museum; pp 426–427 Janine Vangool; pp 428–536 Sharon Forth, Janine Vangool, Linzee Kull McCray; pp 536–543 Janine Vangool; p544 Linzee Kull McCray.

Excerpts from Bagology, How to Save and Sew with Cotton Bags, 1953 Pattern Service For Sewing with Cotton Bags, For Style and Thrift Sew with Cotton Bags, Sew Easy with Cotton Bags, Kasco Home Journal, Thrifty Thrills with Cotton Bags, Smart Sewing with Cotton Bags, Sew Easy with Cotton Bags courtesy Gloria Hall.

Excerpts from A Bag of Tricks for Home Sewing courtesy Amy Barickman.

1952 Pattern Service, Sewing with Cotton Bags, Needle Magic with Cotton Bags, 1954 Idea Book for Sewing With Cotton Bags from the collection of Janine Vangool.

DISTRIBUTOR

To order copies of this book for wholesale, retail or individual sale, please contact:

Martingale
19021 – 120th Ave NE, Suite 102
Bothell, WA 98011-9511 USA

ShopMartingale.com

SECOND PRINTING / PRINTED IN CHINA

©2019 UPPERCASE publishing inc

Every reasonable effort has been made to identify owners of copyright. Reproduction of any company's assets are presented in historical context with no claim of copyright. Errors or omissions will be corrected in subsequent editions. No part of this book may be reproduced in any manner without the written permission of the publisher, except for review purposes.

WRITTEN BY

LINZEE KULL McCRAY

UPPERCASE

FEED SACKS IS VOLUME "F" IN THE UPPERCASE ENCYCLOPEDIA OF INSPIRATION
encyclopediaofinspiration.com uppercasemagazine.com

HISTORY

INTRODUCTION 11
A Bag by Any Other Name 12
Early Flour Advertisements 20

FROM BARREL TO BAG 27
COTTON PICKING FOR COTTON BAGS 28
Burlap: A Cotton Alternative 32
Company Towns 37
From Cotton to Consumer 40
Claude Holsapple, Die Cutter and Designer 42
Designing and Printing Bags 44
Hutchinson Bag Company 61
Biddy the Cat 66
Feed for the Farm 69
SEED SACKS 82
FLOUR SACKS 84
SUGAR SACKS 90
Tradition and Defiance 96

FRUGAL AND FANCY 101
The Habhab Brothers' Rice Sack Shirts 102
Flour Sack Underwear 105

WEARING YOUR BRAND 109
Cotton Bags as Consumer Packages for Farm Products 116

ADDING VALUE FOR WOMEN 123
Bemis Band-Label Patent 126
EMBROIDERY PRINTED SACKS 128
DOLLS 140
Incentives 153

THE GREAT DEPRESSION 156
Cotton Bags Mean Jobs 164
Save the String 168
SOLID COLOUR SACKS 176
Gingham Girl Patent 185

PRETTY PATTERNS 187
   BANDED SACKS 200

WORLD WAR II 213

GLAMOROUS FASHIONS FROM BAGS 223
   Printed Feed Sacks
   Popular as Dresses 230
   Feed Store Wardrobe: Cotton Bags
   Are Useful in Making Things 234
   Beauty Wears a Sack 241
   DRESSES 242
   Feed Sack Competitions 248
   Other Weaves and Fibres 252
   MISPRINTED SACKS 253
   Pullet Polls Predict Elections 260

HOW BAGS WERE USED 263
SELECTING FEED SACKS 265
SHEETS AND BEDDING 270
IN THE KITCHEN 280
   APRONS 282
FUN AND GAMES 294
   Fowl Fashion 299
   DOLL QUILTS 300
CHILDREN'S CLOTHING 302
DRESSES AND SEPARATES
FOR WOMEN AND GIRLS 308
STITCHING ALL THE REST
(EVEN THE UNMENTIONABLES) 313
   LABEL BAGS 314
   PAPER SACKS 324

QUILTS 331
   Feed Sacks Around the World 352

THE END OF AN ERA 353
FEED SACKS TODAY 355
   How to Tell If It's a Feed Sack 358
   Dating Feed Sacks 360
   Washing Feed Sacks 364

SWATCHES

COLLECTIONS 366
   Collecting Companions:
   Gloria Hall and Paul Pugsley 368
   GLORIA HALL COLLECTION......... 370
   PAUL PUGSLEY COLLECTION 392
   Charlene Brewer's
   Feed Sack Notebooks 415
   SCRAPS AND SWATCHES........... 431

BIBLIOGRAPHY AND RESOURCES 541
ACKNOWLEDGEMENTS 543
BIOGRAPHY 544

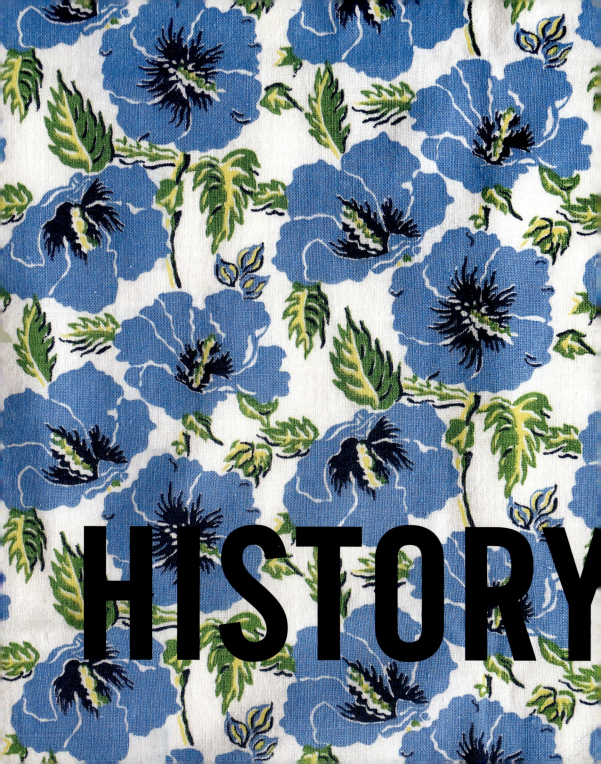

INTRODUCTION

It is easy to appreciate feed sacks—from the bold logos of the earliest printed bags to the mid-century modern prints of the 1950s and 1960s, they offer a seemingly endless array of eye candy for lovers of textiles, typography and design. Their patterns and prints reflect the aesthetic tastes and current events of the times, yet resonate with modern textile designers, who reproduce their compositions or give them a 21st-century twist.

In my first encounter with feed sacks, it was those visuals that drew me in. But that appeal was soon superseded by the story behind the bags, the story of women who saw the possibilities in what was before them—cotton sacks previously filled with fertilizer and feed, and printed with logos of silos, shuckers and sheep—and transformed them into items of warmth and beauty. It is unlikely that those who first turned stiff cotton bags into aprons, diapers, handkerchiefs and dish towels would have chosen the circumstances of geography, economics, race relations and environmental concern that made such reuse necessary. Their "waste not, want not" mentality was essential, whether they lived hand-to-mouth on isolated farms in the late 1800s, sharecropper's cabins in the early 1900s, poverty-stricken towns during the Great Depression or migrant workers' camps after escaping the Dust Bowl. Others may have

An illustration of a woman with a cotton sack from the booklet Smart Sewing with Cotton Bags, *published by the National Cotton Council of America in the 1950s.*

A BAG BY ANY OTHER NAME

Feed sack is the generic name used by many to describe repurposed cloth containers, but the bags go by numerous monikers. Cotton commodity bags is the term preferred by academics, while those who raised poultry called them chicken linen.

For quilters, the term feed sacks (also sometimes written as feedsacks) conjures the florals, stripes and geometric prints that were sewn into curtains, aprons, dresses and quilts. But for decades before (and during and after) these printed bags were available, feed sacks were white and emblazoned with company logos and graphics. Research and first-person interviews have turned up references to feedbags, cotton bags, hen house linen, "pretties," grain bags, seed corn sacks, guano sacks and textile bags, but for ease of reading and writing, unless a piece of cloth is directly identified by its contents, "feed sacks" is the umbrella term used in this volume.

Feed sacks were so named because they originally held animal feed, but cotton sacks also contained everything from flour, sugar, rice and seeds to laxatives and ballots. Cornmeal, bath salts, beans, tobacco, salt, fertilizer, dog food, sausages, iron ore pellets, currency, cement, soap and ammunition were just some of the products distributed in cotton bags.

The vast majority of sewists who used feed sacks to clothe their families and supply their homes with necessary and decorative items were women. Although some men undoubtedly sewed with feed sacks, in this book we refer to all sewists as women.

lived in more abundant circumstances and simply been following the cultural norms of "a penny saved is a penny earned." No matter the circumstances, the clothing, quilts and domestic linens that these women created were often—and often still are—cherished by family members. I want to recognize these women and consider their stories through the objects they made and the materials they had to work with.

Equally intriguing to me is the evolution of cotton bags—from plain white to being printed with patterns for dolls and embroidery to boasting effusive florals, bold plaids and sweet juvenile prints—and the way this demonstrates an industry's recognition of the importance of appealing to wives and mothers, sisters and daughters. As paper bags and other containers gained in popularity, feed sack manufacturers went to great lengths to inspire loyalty to their product, distributing pamphlets with ideas for feed sack sewing, holding competitions with dazzling prizes and offering fabrics created by chic New York designers. Women may not have had a paycheck in hand, but it was clear their opinions were of vital importance in matters of household and farm finances.

Finally, in an era in which throwaway product packaging is ubiquitous, the concept of reusable containers seems like a solution worth revisiting. In a conversation with fabric and quilt designer Denyse Schmidt,

A plaid feed sack with paper band label for Red Hen Poultry Feeds.

The cover of How to Sew and Save with Cotton Bags, *date unknown.*

Denyse applauded the notion of packaging that has a life beyond its original purpose: "If we'd maintained that sense of using up every last bit, instead of this idea of ease and abundance, where packaging is tossed, to some degree we might not be facing the crisis of our environment."

I pay a lot of attention to textiles—it has been the primary focus of the past decade of my writing and editing career. But it was not until 2010 that feed sacks surfaced in my life. Historian and collector Michael Zahs spoke about them at my quilt guild meeting, and I was smitten and wrote about them every chance I got—for Etsy's blog, for the French magazine *Quilt Country* and for UPPERCASE. I learned that the shoe bag I had watched my mom tuck into her suitcase for my entire childhood was stitched from a feed sack. She had selected the grey and turquoise, Mexican-themed fabric while visiting my aunt and uncle's farm. On Craigslist, I found a woman selling a box of feed sacks and could not resist purchasing them—I didn't know exactly what I was getting, but I adored the range of prints, from sweet florals to Atomic Age 1950s design. I could not bring myself to cut into them and they sat on my shelf, a beautiful but unrelenting reminder to do something more with feed sacks. This book is that something more, and when Janine and I decided to work together, four years after first discussing the possibility, I was thrilled. I knew she would do justice to the story and the images, and she has.

The popularity of feed sacks means that there are many books and journal articles written about them; my thanks to the authors who conducted previous research, which was essential in writing this volume. The goal of this book is to provide a broad cultural overview of the life and times of those who used sacks, through research and anecdotal interviews, as well as to share the visual abundance of feed sacks, giving readers an opportunity to appreciate their weave, colour and variety.

My mother's shoe bag, sewn from a feed sack.

From the brochure How to Sew and Save with Cotton Bags.

Workers filling flour sacks at the Sunbonnet Sue flour mill. Photo by Margaret Bourke-White, circa 1939.

Girls in floral dresses, possibly made from feed sacks, peer into a barn.

Even after photographing and viewing thousands of patterns, it is exhilarating to know there are thousands more to be discovered.

There are many ways to tell this story, many angles to take, many details to uncover. Because mills have closed, records have been lost and memories have faded, the history of feed sacks is sometimes difficult to trace. Unlike in the quilting industry today, the majority of those who created feed sack fabrics laboured in obscurity. We have only their thousands of designs by which to remember them.

Fortunately, I *was* able to speak to a number of people about their personal experiences with feed sacks, and I am grateful to those who took the time to share those memories. The feed sack era is receding into the past, making this an opportune time to capture first-person recollections. I was surprised by the many people who have happy memories of feed sacks, as I had read and heard of the associations of feed sacks with poverty. Perhaps because most of the interviewees wore and lived with feed sacks as children (and likely did not do the hard work of preparing and sewing them), the majority looked back on the fabrics with fondness, and admired the dexterity and tenacity of their parents who used them.

And though I did not know their parents, I feel the same way. I stand in awe of women who cooked, cleaned, gardened, canned and raised numerous children (and quite likely, some chickens), and still managed to add rickrack or embroidery to their aprons and dishtowels, who saved scraps until they had enough to make the quilt they envisioned, who stitched a garment that their daughter wore proudly to school. Joyce Stringfellow, who grew up on an Iowa farm with four sisters and a brother, remembers paging through a catalogue to find a dress she liked. "We never bought anything out of it, but my mother would cut a pattern out of newspaper and the fabric out of flour sacks and sew us one like the one in the catalogue," she says. "We had a huge vegetable garden and apple and plum trees and raspberries. I didn't know we were poor. We didn't seem to be lacking anything. Mom was very resourceful."

This book is dedicated to the women like Joyce's mother, whose resourcefulness, resilience and domestic skills sustained and enriched their families and communities, and to the women of today who are finding economic advantage, self-reliance and satisfaction in making do and in making it themselves.

Mrs. Alfred Peterson in Mesa County, Colorado, 1939.

LIBRARY OF CONGRESS

EARLY FLOUR ADVERTISEMENTS

"This Flour always on top," declares Pillsbury's Best in a turn-of-the-century trade card.

Wooden flour barrels were commonplace containers in the late 1800s. This page of advertising is from the Northwestern Miller, 1894.

Washburn Crosby Co.

• OPERATING THE •

C. C. Washburn Flouring Mills

A, B AND C.

Minneapolis, Minn.

PHOENIX MILL CO.,
SUCCESSORS TO
STAMWITZ & SCHOBER.
This Mill Received GOLD MEDAL at World's Fair.
Oldest Firm in Minneapolis.

Flour Excels in STRENGTH and WATER ABSORPTION.

Brands: BEST, PHOENIX, WHITE LILY, VICTORY.

CATARACT MILLS
D. R. BARBER & SON
PROPRIETORS

First Merchant Mill in Minneapolis.
One of the First to Adopt the Patent Process.
Just Thoroughly Remodelled and Enlarged.

Our "WHITE SATIN" POSITIVELY THE BEST FLOUR IN THE MARKET
— CORRESPONDENCE SOLICITED —
D. R. BARBER & SON
MINNEAPOLIS · MINN.

PATENTS: WHITE SATIN, SNOW CRUST
BAKERS: CATARACT, THORN HEDGE

HUMBOLDT MILL COMPANY
MINNEAPOLIS · MINN · U.S.A.
SUCCESSORS TO
HINKLE, GREENLEAF & CO.

PATENT BRANDS: COSMOS, SUPREME, BONANZA
BAKERS BRANDS: CLIMAX, HUMBOLDT, BANNER

DAILY CAPACITY · 1200 BARRELS

L. CHRISTIAN & CO.
OUR MATCHLESS QUALITY
MINNEAPOLIS

This Flour is like its name,
WITHOUT AN EQUAL.
Strong Words but True.

Man's greatest blessing is the light of the sun; the next is
"SUNLIGHT"
—or—
"BEST" flour, made by the NATIONAL MILLING CO., Minneapolis, Minn.
Write for samples.

THE HOLLY FLOURING MILL
MINNEAPOLIS MINN.

FLOUR MANF'RS
Brands
Patent Gold dust
Bakers Inland.
C. McCREEVE Propr.
E. C. PAULL N. E. Ag't. BOSTON.

Various company advertisements from the Northwestern Miller trade publication, 1894.

  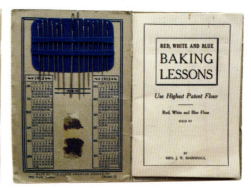

BAGS vs. BARRELS, KEGS, ETC.

The cost of bags is low compared with other types of containers.

Bags as containers tie up a minimum of capital.

When empty, they occupy a very small amount of storage space.

Two bags can be filled in the same time as one barrel.

No special packing is necessary for bags.

They can be closed, sewed or tied in a few seconds.

Trade-marks or advertising in bright inks can be printed on the bags before they are delivered to the user.

They are handled easily and quickly with less labor.

When filled they occupy less storage space than rigid containers.

They give their users a big saving in tare weight.

New, clean bags are attractive and have sales value.

When selected with proper strength they deliver their contents in perfect safety and in good condition.

Consumers may secure the following types of bags: waterproof paper-lined burlap, paper-lined cotton (dirt and sift proof), also bags permitting ventilation and light, bags with crinkled liners, and paper bags for various purposes.

The first cost of rigid, heavy, cumbersome containers is high.

Barrels, boxes, kegs, casks, etc., bought in quantity tie up a large amount of interest-bearing capital.

When empty they use up the same amount of storage space as when filled.

Heavy, expensive containers are cumbersome to handle and require more time and labor to fill.

The rigid containers as a rule require special packing.

They demand much time and labor in nailing on lids and covers and frequently require strap iron.

Rigid containers call for costly labels or stenciling that are expensive to apply.

Because the rigid container is heavy and clumsy more time and labor are required to handle it.

These cumbersome containers occupy more storage space when filled.

Freight rates are high because of more tare weight.

The disposition of empty barrels and kegs presents a problem that involves expense in storage space, freight, handling and other factors.

Your nearest Chase Bag Company Branch or Sales Office will gladly co-operate in helping you solve any bag problem.

The illustration at the left shows two bales of 500 bags each. When filled they will require five railroad cars. Empty, they need but a few cubic feet of storage space. Contrast the space occupied by these thousand bags with the space required for 1000 barrels and the difference in weight between empty bags and barrels.

A Spokane Flour Mills promotion from 1913 included a calendar, needle keeper and baking advice.

This "Bags vs. Barrels" feature was published in the August 1927 edition of Bagology, a trade publication from the Chase Bag Company.

The look of the bag was important to the mill, as demonstrated in this die-cut score card for Voight's Crescent Flour and this "American Special" Russell Milling Company ink blotter.

Losing a Good Customer.

Everything is measured by a standard, whether it be something to eat, to wear or to use. The standard of flour is Pillsbury's Best. No one tries to make any better flour, no one claims that there is any better flour, but some dealers offer for sale other flour which they claim to be just as good as Pillsbury's Best. Why not insist upon having the standard, and avoid the substitute?

Pillsbury's Best Flour

is the best for you and it is the best for those dealers who keep their trade year in and year out by satisfying their customers; but it is not the best flour for those dealers who do not care what they sell, so long as the profit is big. You may be sure that your interests are not thought of when another flour than Pillsbury's Best is recommended.

Pillsbury's Best Flour is for sale by grocers everywhere. Being the best, it is imitated, and consumers are warned not to accept substitutes. "The Best Bread," a book of bread, cake and pastry recipes, sent free.

Pillsbury-Washburn Flour Mills Co., Ltd., Minneapolis, Minn.
Makers of Pillsbury's Oats and Pillsbury's Vitos.

A girl wearing a flour sack costume dress made from Pillsbury's Best flour bags, circa 1910.

MINNESOTA HISTORICAL SOCIETY COLLECTIONS

"No one tries to make any better flour, no one claims there is any better flour, but some dealers offer for sale other flour which they claim to be just as good..." Pillsbury Washburn Best Flour Mills, Minneapolis, Minnesota, 1900.

A Victorian trade card for Wonder Flour, Lake Superior Mills.

"A Good Loaf, after a well made loaf." This die-cut recipe card from the 1890s opens to reveal an advertisement and recipe from the Washburn Crosby Company, using Gold Medal Flour.

FROM BARREL TO BAG

For centuries, boxes, tins and barrels were used to transport and store all kinds of goods, from seeds to gunpowder. Although sturdy, barrels were heavy—both when empty and even more so when full. Boxes and barrels were also awkward to lift and move, and they became more expensive to produce as first-growth forests dwindled. But until the invention of the sewing machine, alternatives were slim.

Textile bags had advantages: they were more easily thrown over a horse's back or man's shoulder than a barrel, and could be stacked on top of one another, allowing more goods to be stored in a smaller footprint. But early bags were largely hand-woven and hand-sewn, and therefore expensive. Their scarcity meant they were used repeatedly and often printed or embroidered with the owner's name or initials, ensuring their return from a mill, where they might have been brought full of wheat or corn.

The invention of the sewing machine changed all that. Inventors had been tweaking the sewing machine since the late 1700s, but by the mid-1840s a reliable lockstitch machine had been developed. Bags could now be sewn with a stitch that created strong side seams with a top seam that opened when someone undid the stitches at either end

Elias Howe designed this lockstitch sewing machine in 1845. It was the most practical of several early inventions.

of the bag and pulled out the line of stitching. The Chase Bag Company, founded in Boston in 1847 by Henry Chase, employed seamstresses to stitch bags on treadle sewing machines, one at a time. In 1849, Chase produced a machine-made, chain-stitched flour bag in collaboration with inventor John Batchelder, who patented an adaptation for sewing machines that enabled cloth to be continuously fed under the needle. In 1864, a machine for creating large quantities of bags arrived in time to help fill the demand created by the American Civil War, when bags were used for everything from transporting food and medical supplies to fortifying military positions with piles of sand-filled bags.

In addition to improved sewing technology, the availability of cheap cotton helped fuel the move from barrels to bags. Cotton was a tremendously important crop in the United States, both for stateside use and as an export. Cotton remained inexpensive even though it was picked by hand, in part because labour costs were rock-bottom: labour for field work and cotton picking fell first to slaves in the southern United States, and then to sharecroppers after the Civil War all the way through the 1940s. As late as 1950, cotton pickers were paid just three cents a pound for their labours. Though a mechanized cotton picker appeared in 1936, it was not until the 1940s that large-scale machine picking became practical. Cotton also had to compete with fibres like rayon, which was in great favour during the Depression for undergarments and other clothing, and this competition also kept prices low.

COTTON PICKING FOR COTTON BAGS

Cotton pickers carried bags known as pick sacks into the fields. They featured an attached shoulder strap, and varied in size from six-foot bags used by children to bags nearly 12 feet long that were dragged down the furrows of a field as the picker filled them with cotton. These bags were sometimes stitched from repurposed feed sacks. Some manufactured pick sacks were reinforced with a layer of asphalt, and later with plastic, to reduce the wear from being dragged.

Lain Adkins picked cotton on his grandmother's Oklahoma farm in the mid-1950s. He started picking when he was seven years old and is matter of fact about what seems to be a difficult task for a small child. "It was a lot easier for me to pick in a way because I was right down there with it—those big, tall folks had to bend over," he says. "And I was a kid,

Cotton pickers in the Southern San Joaquin Valley, California, photographed by Dorothea Lange, 1936.

Boys picking cotton in Pulaski County, Arkansas, 1935. Photo by Ben Shahn.

Loading cotton. Southern San Joaquin Valley, California, photographed by Dorothea Lange, 1936.

"She's a good steady picker. Works all day long."

Edith, a five-year-old cotton picker in Texas, photographed by Lewis Wickes Hines in 1913.

Cotton pickers in Lehi, Arkansas, photographed by Lee Russell in September 1938.

LIBRARY OF CONGRESS

A woman, possibly wearing a dress made from a feed sack, drags a bag of cotton, in September 1938 at the Lake Dick Project in Arkansas. Photo by Lee Russell.
LIBRARY OF CONGRESS

so I had a short sack and wasn't dragging as much weight around. Picking the cotton isn't the hard part; dragging the sack behind you is."

Lain's off-white sack was stitched from other bags, "probably from feed sacks—we called them tow sacks," and he remembers the shoulder strap as being two to three inches wide and not particularly uncomfortable. Lain and the other workers walked between rows of plants, picking cotton from both sides while the bag dragged behind in the furrow between the rows. When their sacks were full, they would haul them to a wagon parked in the field that was fitted with a scale. The sacks were weighed, the weight was recorded and the bag emptied—then the process started again. "I earned a few pennies and that was kind of fun," he says. "But no kid, after an hour or so, wants to continue doing something dull and boring."

BURLAP: A COTTON ALTERNATIVE

During the American Civil War, the South restricted the North's access to cotton, and above the Mason–Dixon line, burlap bags replaced the more finely woven cotton bags. Burlap is a sturdy, plain weave fabric created from jute, which was imported from India. The rough fibres of burlap bags meant they were recycled for mostly outdoor purposes, though they sometimes served as a canvas for a hooked rug or for the underside of the seat of an upholstered chair. After the Civil War, burlap bags continued to be used for some animal feeds and other agricultural products, until 1942, when all burlap was requisitioned for military use during World War II. Today, burlap bags are filled with potatoes, tobacco, coffee, nuts and other products.

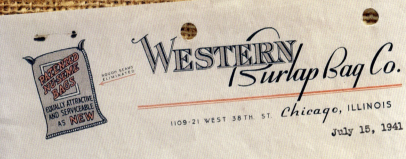

PATENTED NU-SEME BAGS
EQUALLY ATTRACTIVE AND SERVICEABLE AS NEW

ROUGH SEAMS ELIMINATED

WESTERN Burlap Bag Co.

1109-21 WEST 38TH. ST. Chicago, ILLINOIS

PRINTED NU-SEME BAGS
CUT SACKING COSTS

July 15, 1941

Central San Francisco
Guayanilla, Puerto Rico

Gentlemen:

Because of a decided shortage in burlap and Heavy Twill bags, it is quite possible that you might be interested in Heavy Twill bags of a similar size as that which you have been using for raw sugar in the past.

At this moment, we can offer about 100,000 of these bags measuring 26 and 27 inches wide and 45 inches long at approximately 19½¢ per bag, FOB New York City or New Orleans, or 17¾¢ per bag, FOB Chicago.

If interested in these bags, we would be pleased to submit sample for your inspection and our offer, of course, is made subject to further market changes and our confirmation.

Yours very truly

WESTERN BURLAP BAG COMPANY

D. R. Goodman

Secretary

DRG/ar

Feed and flour mills typically bought bags to hold their products from bag manufacturing companies, of which there were many. Some of the largest and best known included Chase (which started in 1847 in Boston and expanded to include plants in New Orleans, Toledo, Philadelphia, Milwaukee, Cleveland and Memphis), Bemis (founded as J. M. Bemis and Company in St. Louis in 1858 and today headquartered in Minneapolis—its plants included sites in San Francisco, Seattle, Winnipeg, New Orleans and Indianapolis), Werthan Industries (founded in Nashville in 1900 and today operating as Werthan Packaging in White House, Tennessee), Fulton Bag Company (founded in 1868 in Atlanta, Georgia, and operating as part of Fulton Industries and Allied Products until the Atlanta mill closed in 1978) and Percy Kent (founded in Brooklyn, New York, in 1885 and acquired by Gateway Packaging in 2000, operating today out of St. Louis). By 1946 there were more than 30 such bag manufacturing companies, and bags accounted for 4.5 percent of total United States cotton consumption.

Other manufacturers included Laurel Mills of Laurel, Mississippi; Flint River Cotton Mills of Albany, Georgia; Lone Star Cotton Mills of El Paso, Texas; Illinois State Penitentiary of Joliet, Illinois; Cannon Mills Company of Kannapolis, North Carolina; and Royal River Mills of Yarmouth, Maine.

Bag companies created their products with woven goods purchased from textile mills. Some bag manufacturers, including Bemis and Fulton, took over the entire process, from preparing the cotton to spinning fibre and weaving fabric, to manufacturing the bags, designing their exteriors, and selling them to the feed and seed dealers, flour mills and others who filled them with their products.

This paper sticker was affixed to a letter from the Burros Bag Company, November 15, 1941, Brooklyn, New York.

Letterheads for Lone Star Bag, Arkell Safety Bag, Burros Bag, Western Burlap Bag and Chase Bag companies, 1940s.

Various bag designs illustrated on a Stanard Tilton Milling Company invoice from St. Louis, Missouri, from the 1920s.

LONE STAR BAG & BAGGING CO., INC.

QUARTER MILLION DOLLARS
CAPITAL STOCK PAID IN

BARTLETT ARKELL, President
EDWIN D. GREENE, Vice Pres't & Treas.
M. M. YELD

ARKELL SAFETY BAG COMPANY

MANUFACTURERS OF

"ARKSAFE" ELASTIC PAPER LININGS FOR BAGS, BARRELS DRUMS AND BOXES, SHIPPING BAGS AND ...
WATERPROOF LIN...

"ARKSAFE" SHIPPING BAG ELASTIC LINING PATENTED NOV 19TH 1901

"ARKSAFE"

Cable Address "ADV...
CABLE "BURROSBAG" NEW YORK

TELEPHONES: MAIN 4-2424-5-6   "IT'S IN THE B... NECT"

BURROS BAG COMPANY, INC.

BAGS · BAGGING · BURLAP · NEW & SECOND HAND
...FACTURERS · DEALERS · IMPORTERS · EXPORTERS

PLYMOUTH, ADAMS & JOHN STS.
BROOKLYN, NEW YORK

November 15, 1941

PATENTED NU-SEME BAGS — ROUGH SEAMS ELIMINATED
EQUALLY ATTRACTIVE AND SERVICEABLE AS NEW

WESTERN Burlap Bag Co.

1109-21 WEST 38TH ST.   Chicago, ILLINOIS

PRINTED NU-SEME BAGS
CUT SACKING COSTS

"Bags of all Kinds"

CHASE BAG CO.

(ESTABLECIDA EN 1847)

P.O. DRAWER 1590
NEW ORLEANS, LA.
E.U.A.

DIRECCION TELEGRAFICA CHASEBAGS
CLAVE | A.B.C. 5ª EDICION
      | BENTLEYS
      | ACME

35

Cotton bags were the perfect surface for printing on, and logos began to appear on them in the 1880s. Initially, many logos were round, their iconography taken directly from the top of the barrels previously used by feed, seed and flour companies. Over time, labels and illustrations expanded to take better advantage of the rectangular shape of the bags, and included everything from champion corn shuckers to Western heroes to illustrations of the animals that were to be fed by the contents. Exactly who designed these sometimes bold, sometimes sweet logos is largely lost to history, though a few stories remain.

Elegant art from the front of an envelope for the Springfield Flour Mills, above, with additional seals shown on the back (enlarged for detail).

COMPANY TOWNS

MADISON COUNTY ARCHIVES

Around the turn of the century, the Bemis Bag Company purchased 300 acres in cotton-growing country in Tennessee and created the planned community of Bemis. The town's centrepiece was a textile mill that produced fabric and thread used to create bags at the Bemis plants. According to company history, the town offered employees "all the necessities of life at the turn of the century, including schools, churches, shops and even a YMCA." Wide streets, ample landscaping, a swimming pool and an 850-seat auditorium were built in the next decades. In the mid-1960s, employees were given the opportunity to purchase their own homes as Bemis began selling off parts of the town. Bemis operated the Jackson Fibre Department and Cotton Mill until 1980, and today the town, which is on the National Register of Historic Places, is part of the city of Jackson. In 1928, Bemis established another company town, Bemiston, Alabama.

View in Bemis, Tenn. The finest cotton mill village in Southern States

AMERICAN TEXTILE HISTORY MUSEUM, LOWELL, MA, USA

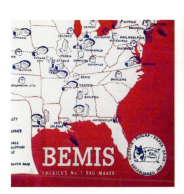

Bemis, Tenn. Section of the new part of the village

37

ADVERTISING SPACE

...NOT FOR SALE

PLACE all the bags you use in a year side by side and you would have the equivalent in display space of a gigantic billboard.

Then multiply each unit by the many individuals who see your name and trademark before and after the bag reaches its destination. You now begin to see the surface of your bags as an advertising medium worth a fortune in terms of paid space — space you wouldn't sell for love or money.

That's the way smart merchandisers view the square feet of "frontage" their bags represent. That's why so many appreciate the sharp, brilliant, eye-catching impressions on Chase bags.

Often, too, Chase bag designers are able to help increase the "eye appeal" of each bag before it joins the parade of products that is viewed each year by countless millions.

CHASE

BURLAP AND HEAVY DUTY COTTON BAGS

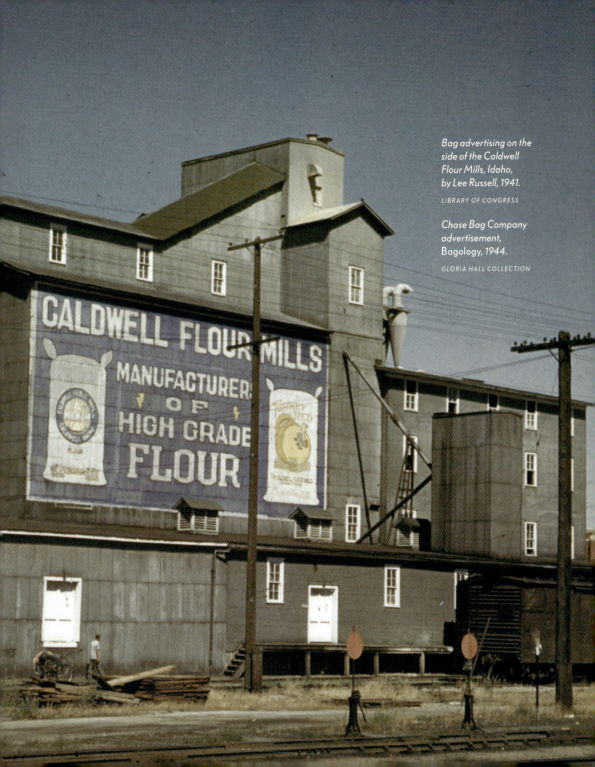

Bag advertising on the side of the Caldwell Flour Mills, Idaho, by Lee Russell, 1941.
LIBRARY OF CONGRESS

Chase Bag Company advertisement, Bagology, 1944.
GLORIA HALL COLLECTION

FROM COTTON TO CONSUMER

The journey from cotton to cotton bags started in the field, where cotton was picked, weighed and then sent to market. Cotton was ginned, a process in which the seeds are separated from the fibre, then carded, a procedure that aligns the fibres directionally. This carded fibre was spun into thread and yarn, and used to weave the fabric for the bags. These activities were done by hand in the case of the earliest cotton bags; the work became much less time consuming once mechanized. Most cotton was grown in the South, and New England was a major site of textile manufacturing.

The yardage woven at textile mills was sold to bag manufacturers (although some manufacturers, including Bemis and Fulton, had their own textile mills) who cut and sewed the bags and designed and printed the labels. They were then sold to companies who that needed them to ship their products. These companies filled the bags with goods and sold them through flour mills, general stores, grocery stores, implement stores and other feed and seed dealers—wherever the items they held were sold.

Photographs of the Fulton Bag and Cotton Mills of Atlanta, showing threading and yarn spinning machinery, as well as sewing machines circa 1910–1930.

GEORGIA TECH ARCHIVES

CLAUDE HOLSAPPLE
DIE CUTTER AND DESIGNER

Claude Holsapple practicing his craft, circa 1929.

Framed bags are two examples of Claude Holsapple's artwork for the Werthan Bag Company. Watermill Flour was printed in 1934 and White Way Flour in 1929.

This miniature promotion advertised the Werthan company's Double Duty Bag and could be used as a pin cushion.

RICKRACK.COM

From the time she was 18 years old, Julie Hardgrave worked at her grandfather's engraving business, Claude L. Holsapple & Son, in Dallas, Texas. In the early years of her employment, her desk was close to her grandfather's, and though Claude was a quiet man, through conversations she learned about bits and pieces of his life. One day in the mid-1970s he was rummaging in his desk and pulled out a brown paper sack. He asked Julie if she had ever seen what was inside. She hadn't. Her grandfather proceeded to unroll five flour sacks and tell her the story behind them.

When he was just 10, Claude's mother died and his father sent Claude and his brother to live on a tobacco farm with grandparents they barely knew. Although Claude refrained from telling Julie about this period, she is sure it was not a happy time, and Claude left the tobacco farm just four years later. He had little education and possessed few sellable skills. Around 1922 or 1923, when he was 17 or 18, Claude took a janitorial job at the Werthan Bag Company in Nashville, where he became fascinated with the exacting work of the die cutters. Die cutters created imagery on the bags by painstakingly carving away bits of metal, leaving raised areas that were used to imprint logos and text. Illustrations and words had to be drawn in mirror image, each colour on a bag required a separate die, and all the dies had to line up in order to create a sharp registration on the sacks. Claude taught himself these skills, thanks to a die cutter who noticed Claude's interest and loaned him a book and tools. "Gramp was evidently quite talented, because by the late 1920s he was the head die cutter," says Julie.

The bags her grandfather showed her that day were solely his designs—Julie believes that designs were typically created by more than one person—and ones he had pulled off the production line, rolled up and saved. Julie never learned from her grandfather how designs were planned, but she assumes that clients had input, given that each of the bags he saved depicts a very different scene (one of these was a Blue Boy flour bag, and another a bag with a portrait of George Washington.) Julie marvels at the detail and true colours of the designs. Claude told her it was illegal to mark the dies, but that he had worked his initials, CH, into the each of the designs, though he could not remember exactly where. Each bag had its production date written across the top by Claude, in pencil.

Julie's grandfather gave her those flour sacks that day, and she has since had them framed and has shared them with other family members. Though Julie loves the bags, it is a photo Claude also gave her that day of him bent over his work that she especially cherishes. "We had zero photos of my grandfather before my grandmother came along, so I was just blown away," she says. "Though he was proud of his designs, he told me cutting dies was tedious work and you can see it in the picture. It's hard on your neck, because you're looking down all day long, mostly through a magnifier."

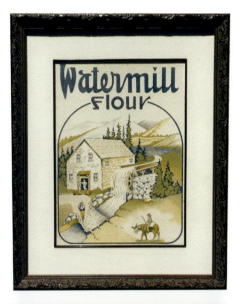

In 1935, the Fulton Bag Company offered to make Claude head of their die-cutting operations in either Minneapolis or Dallas, and because he did not like cold weather, he chose Dallas. He left Fulton in 1943 to open Holsapple Engravers. (Claude told Julie that engraving was "a walk in the park" compared to die cutting.) Claude continued engraving until he was 80, in 1985, and passed away in 1990, handing down the business to his only son, Julie's father. Today, it is run by Julie and her brother and sister. The engraving business has changed significantly over the years. Claude taught the art of hand engraving to numerous students, but because it is so time consuming and expensive, few engage in it today. Julie's brother does most the shop's engraving by machine.

Though the skill set needed for engraving has changed, Julie is grateful for the years she sat alongside her grandfather and for the opportunity to see the work he did in the past. "I study the images on those bags sometimes and try to figure out how he did that—it fascinates me," she says. "I was just lucky, sitting down by him every day."

DESIGNING AND PRINTING BAGS

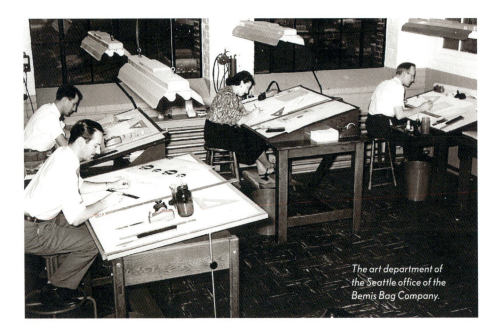

The art department of the Seattle office of the Bemis Bag Company.

Bag companies bought cotton yardage from textile mills to create their products. Once sewn, bags were stamped with small trademarks identifying their maker (Biddy the cat, for example, can be found on bags made by Bemis—such logos are frequently found near the tops of bags). Flour and grain mills, feed and seed companies, and others placed orders for these bags and worked with the bag companies to create labels. While a very few of these designers have been identified (see pages 42-43), most remain anonymous—some may have been professional artists and designers, while others were talented amateurs. In her book *Soft Covers for Hard Times: Quiltmaking and the Great Depression*, author Merikay Waldvogel writes that Albert Werthan of the Werthan Bag Company hired "one woman artist" to create designs, and that the company had in-house intaglio printworks in 1939. She also notes that Bobbie Durrett, owner of the Ringgold Mill near Clarksville, Tennessee, reported that salesmen from three bag companies visited his flour mill frequently and helped develop trademarks for the bags. Once the bags were completed and delivered, they were filled with the appropriate product, sewn shut and sold to consumers directly or through stores.

Dress print bags were created in a similar manner, although rather than being printed directly on the bags, most labels were adhered with paste and/or sewn into a seam, making the bags easier to reuse. In some cases, the fabric woven by textile mills and sold to bag manufacturers was the very same fabric woven for stores to sell by the yard.

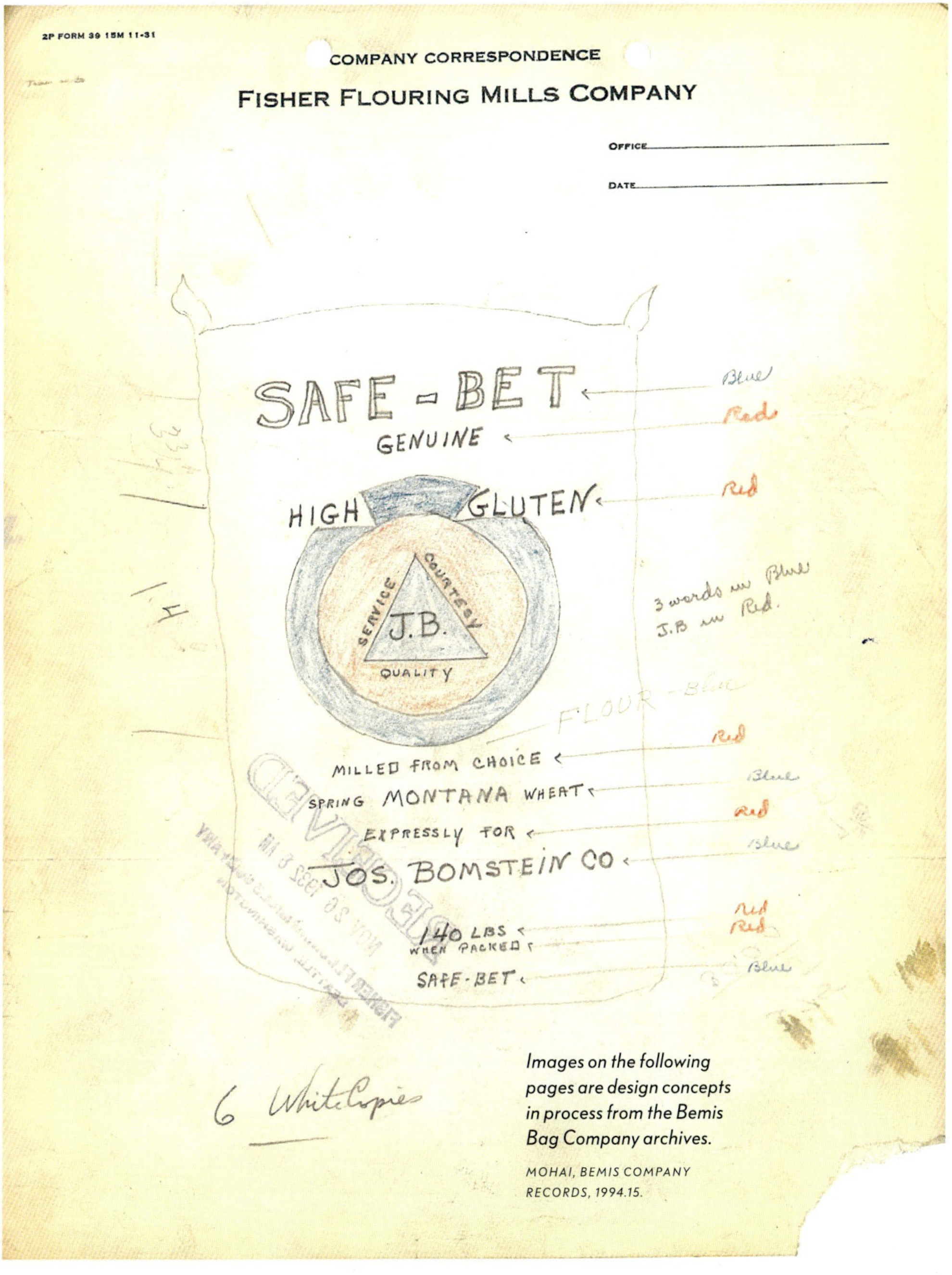

2P FORM 39 15M 11-31

COMPANY CORRESPONDENCE
FISHER FLOURING MILLS COMPANY

OFFICE_____

DATE_____

SAFE-BET ← Blue
GENUINE ← Red

HIGH GLUTEN ← Red

SERVICE COURTESY
J.B.
QUALITY

3 words in Blue
J.B. in Red.

FLOUR - Blue

MILLED FROM CHOICE ← Red
SPRING MONTANA WHEAT ← Blue
 ← Red
EXPRESSLY FOR ← Blue
JOS. BOMSTEIN CO. ←
140 LBS ← Red
WHEN PACKED ← Red
SAFE-BET ← Blue

6 White Copies

Images on the following pages are design concepts in process from the Bemis Bag Company archives.

MOHAI, BEMIS COMPANY RECORDS, 1994.15.

This illustration from the Bemis Bag Company depicts the life of a dress print bag—from the desk of "one of America's leading artists" to the "scores of things" made with the fabric at home.

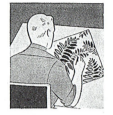

Print the cloth with colorful, attractive patterns. (Bemis' exclusive patterns are designed by one of America's leading artists.)

Make the cloth into bags for feed and flour. (Bemis, a leader in the field, makes millions of printed cloth bags every year.)

Distribute that feed and flour to American farms and families. (Bags carry practically all of the nation's feed and flour to market.)

Make the printed cloth from the bags into dresses, aprons, curtains . . . scores of things. (So the cloth does double work . . . *for a long time.*)

A man at work in the art department of the Seattle headquarters of the Bemis Bag Company. Date unknown.

MOHAI, BEMIS COMPANY RECORDS, 1994.15.

The "Genius" artwork sketch at left would have been placed beneath the open frame in this rough design to show how the illustration and text would appear on the flour bag.

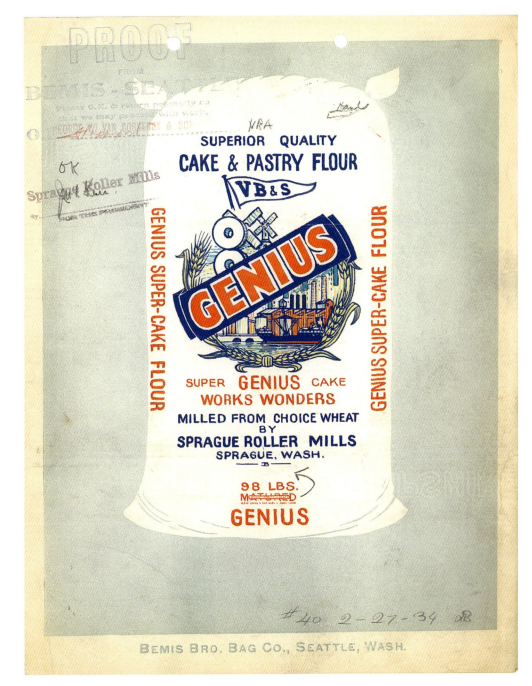

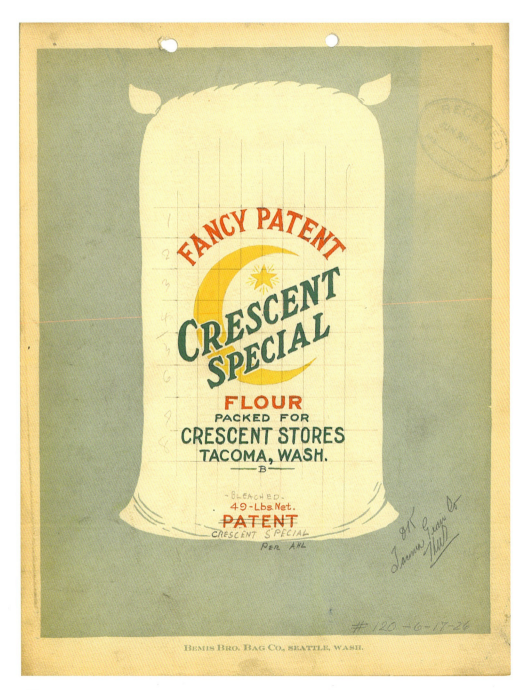

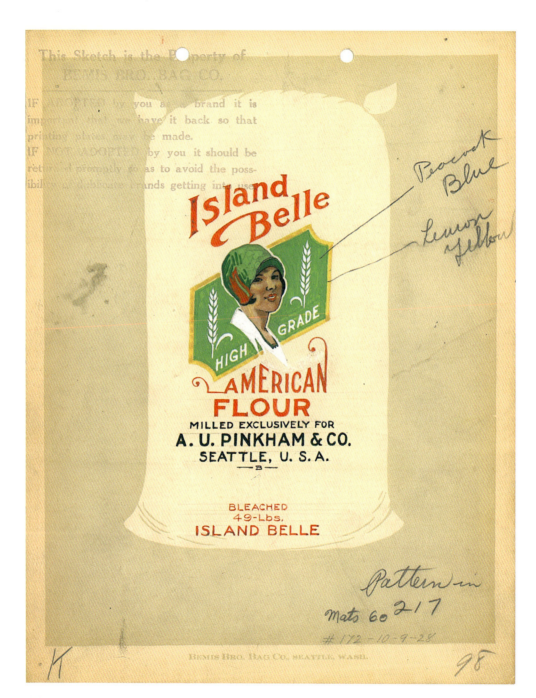

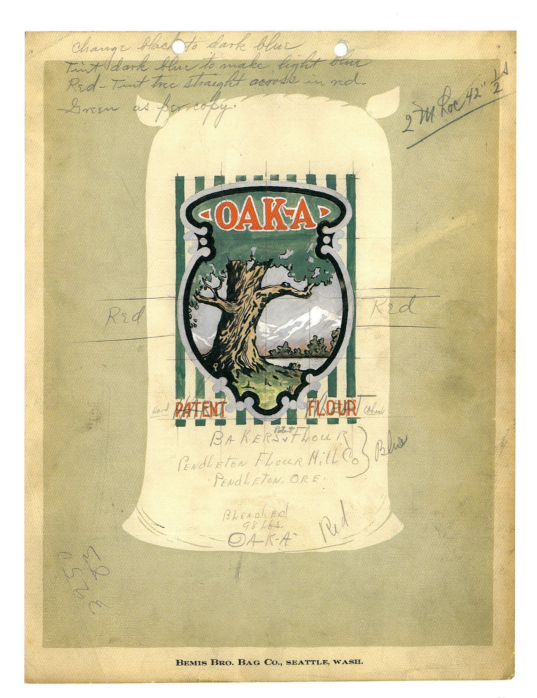

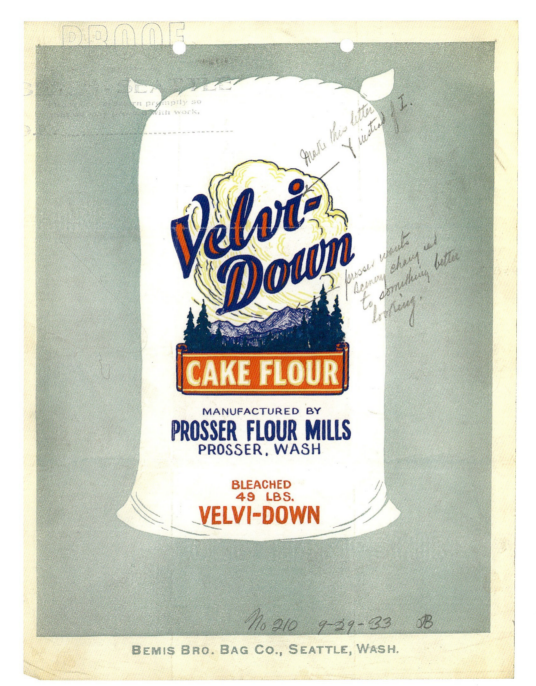

Published in the October 1927 edition of the company magazine, Bagology, *this photographic feature tours the Chase Bag Company facilities: the engraving room, sewing room, stereotyping department, loading platform, press room and the finishing operations.*

Two-colour web presses from the Seattle factory of the Bemis Bag Company used to print burlap bags.

MOHAI, BEMIS COMPANY RECORDS, 1994.15.

Chase Bag Company **BAGOLOGY** *Page Three*

Engraving, routing and matrix making.

A section of the sewing room.

The stereotyping department.

Spacious loading platform in the new quarters.

These automatic presses print, cut and fold the bags.

"Turning the bags"— the finishing operations in bag making.

FOR LETTER SEATTLE 2 color
Web Press

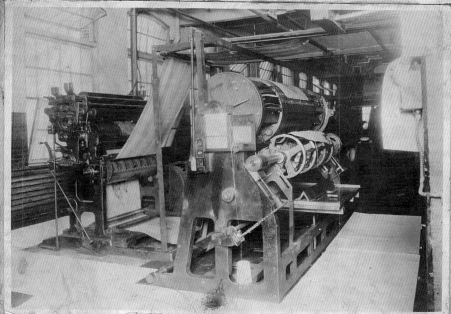

Bag Printing is Difficult Art

THE printing of trade-marks, brands, names and addresses on the many kinds of Chase Bags is a most difficult art.

We wish that you could step into the printing departments in our different factories and watch them at work. In one of our branch plants, you would see a battery of presses printing ten to fifteen thousand bags a day, in several colors, simultaneously on burlap bags.

At another plant you would find the department at work printing beautiful many-colored trade-marks on paper-lined cotton. At still another factory the presses would be printing paper-lined burlap bags and bags of Saxolin Open-Mesh cloth. At another Chase Bag Co. plant, a battery of large, four-color, high-speed presses is at work printing bright colored inks on brilliant, white-coated, polished paper sacks.

Printing on these various kinds of materials requires infinite knowledge of the printing art. It requires concise information on the making of many kinds of engraved plates, one for each color, which will register correctly. It calls for a knowledge of inks best suited to the particular material, and above all, it necessitates the employment of highly skilled craftsmen, working under the watchful supervision of Chase Executives.

To print clear-cut, brightly-colored brands of the many types of Chase Bags, is an art—the result of 81 years of study on bag printing.

Bagology, *June 1928.*

Bemis Bag Company.
MOHAI, BEMIS COMPANY RECORDS, 1994.15.

HUTCHINSON BAG COMPANY

The Hutchinson Bag Company (known as Hubco since 1991), has been in the textile packaging business for nearly a century. Founded by Emerson Carey and James Dick in 1919 in Hutchinson, Kansas, the company started manufacturing bags for salt. In response to market needs and acquisitions of other companies over the decades, Hubco currently makes bags in cotton, burlap and synthetic and non-woven materials for mining and oil companies. They have also maintained a niche in food and specialty packaging such as cotton bags for dry food mixes, flour and rice, as well as packaging for sausage, ham and poultry bags.

Fabric is purchased on rolls from other textile companies; splitters on site cut it down to appropriate sizes. When it comes to printing graphics on plain cotton sacks, the process has remained relatively unchanged for decades: letterpress equipment from the late 1940s and 1950s is used to print up to four spot colours. The Hubco sewing department stitches them into bags.

Now a unique specialty, Hubco still offers a selection of dress print flour sacks. Books for readers who supported this project through pre-orders were packaged in a contemporary Hubco dress print sack.

"Our operations have been in the same plant since 1919," says Lyle Smart, a sales representative at Hubco. "We have removed the ivy from the exterior, closed in the shipping dock and added onto the back of the building." As other companies have gone out of business over the decades and because competitors sell imported bags, Hubco has kept their manufacturing within the United States—contributing to their longevity and proud heritage.

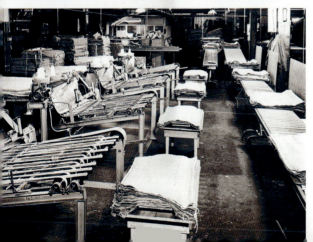

PHOTOS COURTESY HUBCO. PHOTO OF WOMEN SEWING COURTESY THE KANSAS STATE HISTORICAL SOCIETY.

Ads that appeared in Bagology in the 1940s include some stereotypical depictions of women.

The American Finishing Company's fabric sample for the Bemis Brothers Bag Company, 1963.

The Chase Bag Company coined the term "bagologist" to describe the salesman informed on all manner of bag manufacturing (Bagology, June 1928).

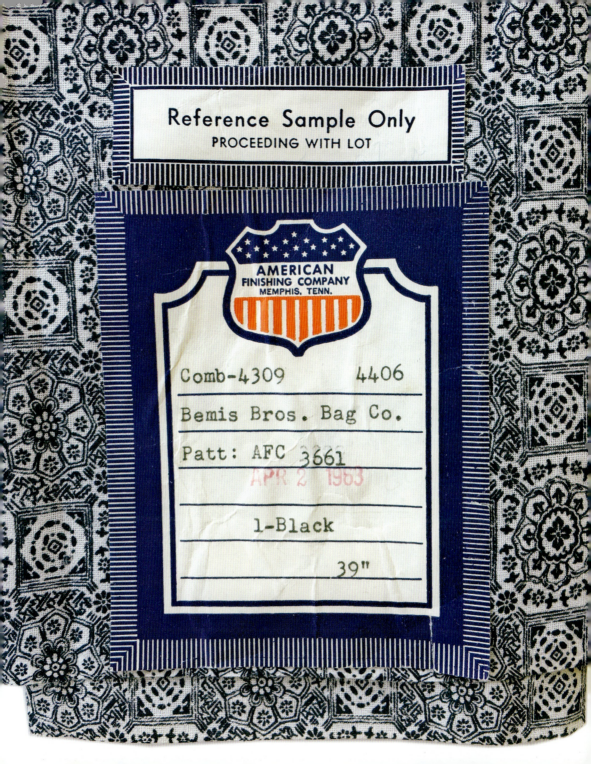

31145

BIDDY THE CAT

The most memorable (and perhaps the most adorable) logo found on cotton bags was Biddy the cat, the mascot of the Bemis Brothers Bag Company. The story goes that Biddy was a real cat and an excellent mouser in Bemis' first factory in St. Louis. Judson Moss Bemis, the company founder, decided in 1881 to use the image of Biddy emerging from a sack as a way of saying he had nothing to hide—that he was "letting the cat out of the bag" and would deal honestly and fairly with his customers.

The company opened its second branch in Omaha in 1887, and likenesses of Biddy decorated the building. Photographs from 1898 show several sculptures on its exterior featuring cats in bags. (The building has since been demolished.)

In addition to serving as the subject of sculpture and having her image printed on bags, Biddy appeared in company communications and newsletters. In 1894, Biddy was shown with her four kittens, who represented the company's four branch factories in Minneapolis, Omaha, New Orleans and West Superior, Wisconsin. In 1920 she was depicted playing bagpipes in Minneapolis and baseball in Omaha, and in 1921 she sheltered from a storm beneath an umbrella and announced the May Day dance in Seattle. Biddy was redrawn over the years—and depicted both in and out of the bag—until she was finally retired in 1962.

HIS FLOUR ABSOLUTELY GUARANT

GET RESULTS "HIT THE BULL'S EYE"

BULL'S EYE

WITH COD LIVER OIL

GROWING MASH

AND BUTTERMILK

HOLDREGE ROLLER MILLS

HOLDREGE, NEBR.

FEED FOR THE FARM

The largest user of cotton commodity bags was the flour industry, according to researcher Gloria Hall, who says that 52 percent of what we term "feed sacks" were actually flour bags. Still, it is no accident that we refer to these bags as "feed sacks," given that animal feed was produced in every American state and filled about 12 percent of cotton sacks. Empty, the sacks were a valuable commodity, frequently used for sewing women's clothing—it took three or four to make a typical dress. While many of those interviewed for this book said their families were self-sustaining, most if not all used animal feed to supplement the diets of the animals on their farms.

A stable food supply is critical to farmers and ranchers who raise animals for sale, and the practice of feeding domesticated animals is thousands of years old. But in the early 19th century, scientists began measuring the nutritional value of food like hay, and by the late 1880s feeding standards were established. The byproducts of grain milled for human consumption were originally dumped in waterways, but when the government expressed concern about the environmental effects of doing this, the byproducts were instead fed to animals, and the feed industry was born. Cities like Minneapolis and Chicago, which were already the sites of myriad mills, soon became home to large feed operations.

In the 1920s, Purina was the first company to produce animal feed in pellet form, enabling a blend of nutritional ingredients to be more easily balanced and delivered to livestock. Compressing the various ingredients into a pellet also prevented animals from picking and choosing at mealtime, ensuring that they would eat all the nutrients in the blend. In the decade that followed, the pellet became the norm for animal feed, and hundreds of smaller mills expanded beyond the major milling cities to locations closer to farms. Feed sacks reflect that move, and the logos printed on them are of mills in larger cities like Denver and Fort Worth, as well as smaller towns like Beatrice,

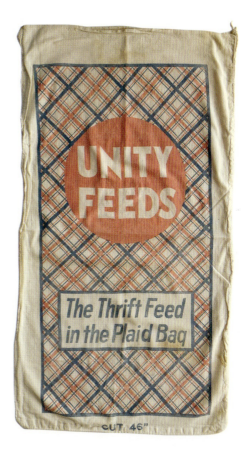

Some feed bag designs were simply printed in few colours, but used bold graphics to catch consumers' eyes.

From chicks to hogs to cows—a graphic sampling of feed sack artwork.

Nebraska; Chippewa Falls, Wisconsin; and Kutztown, Pennsylvania. Some of the most delightful labels also give an indication of the animals farmers were feeding—chickens, pigs, cattle, horses, mules and sheep.

It was—and still is—recommended to give young animals food to enhance their growth. Baby chicks, for example, eat "mash" and "starter mash" until they are four to eight weeks old. Mash includes several ingredients and has been ground into small pieces, making it easier for young birds to eat. Starter mash includes ingredients specifically for growth. Food for older chickens includes blends for "broilers" (to encourage fast growth for maximum profit) and "layers" (containing extra calcium and protein for egg-laying poultry). Food for older animals is typically in pellet form. All this information could be found on the labels on "chicken linen" and on sacks containing food for pigs, cows and other animals.

In the August 1947 issue of *Bagology*, a Chase Bag Company publication, an article saluting the feed industry notes that it had kept up with increasing demands for more food: "The rate of increased production of eggs, milk, and meat, which has been accomplished through improved feeding in the past 35 years, is truly astounding. … In 1912, hens layed an average of about 70 eggs a year. Today that average has more than doubled. … With improved feeding a cow will produce 150 percent more butterfat for approximately 20 percent additional feed cost. The Feed Industry is also credited with developing methods of accelerating the growth and improving the quality of meat stock." The article suggests that the feed industry's motto of "service to agriculture" was too modest, and instead should be "service to all mankind."

As farms grew in size, larger feed mills became the norm and buying feed in bulk became more cost effective. Even for smaller quantities, feed sacks were used less often, and other materials replaced cotton bags. Today, feed sacks are typically made of polypropylene plastic.

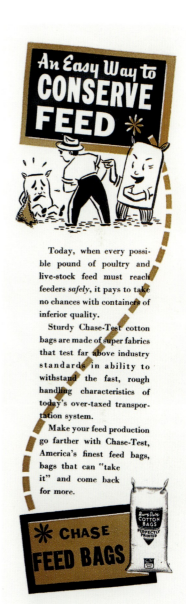

A farmer in front of a seed and feed store in North Dakota, July 1936. Photo by Arthur Rothstein for the Farm Security Administration.

LIBRARY OF CONGRESS

Various matchbook covers advertising feed and seed dealers Rogers Better Range Feeds of Ainsworth and Abbotstown Grain and Feed Store as well as the Foltz Flour Mill of Salem, Ohio.

This 1949 advertisement for Nunn-Better All Mash Poultry Feed highlights the dress print bag packaging.

Printed in blue, turquoise and red inks, this eye-catching Hubbard's Sunshine Egg Concentrate bag uses overprinting to achieve the perception of additional colours.

Bemis "A" Extra Heavy Seamless were durable bags designed to pack and store agricultural seed. "Because of the strength that results from weaving more threads to the inch, they are extra tough, non-sifting, and non-raveling. They'll safeguard your seeds from harvest time to planting time."

These bags reflect a range of colours and designs found on seed corn, wheat and sorghum sacks.

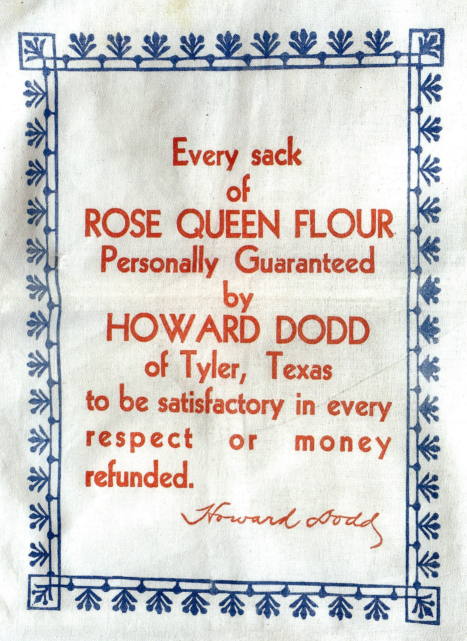

From Lucky Boy to Queen, flour bag artwork was designed to appeal to homemakers. Collector Paul Pugsley says that the "Queen" depicted is child actress Shirley Temple.

Salty or sweet? These bags once held cane sugar and table salt.

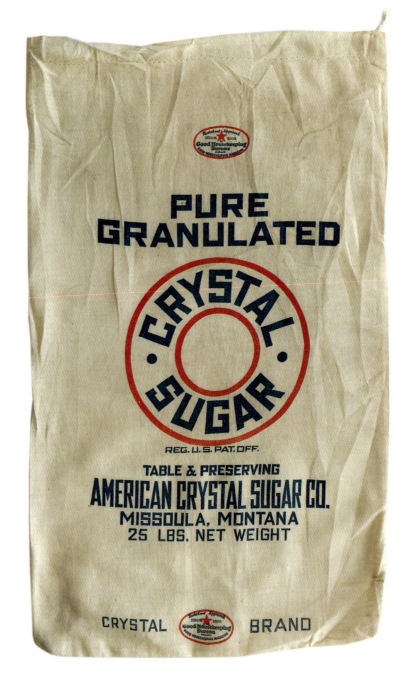

C and H Menu Sugar ad from 1932.

Advantages of Cotton Bags for Flour

FOR all around service, a cotton bag is unquestionably the logical container for flour as evidenced by the fact that the largest proportion of flour is today and has been for almost 100 years shipped in cotton bags. Here are a few advantages that have made cotton bags predominate for shipping flour.

1. Cotton flour bags are built specifically for shipping this commodity. From the selection of the cotton to the completion of the bag, every step in the process of manufacture is governed by years of experience with this type of container.

2. Unusual care in selecting the cotton, the weaving of the cloth and the designing of the cotton bag insure greater strength.

3. Strength in turn gives safety through less breakage, better arrivals and fewer damage claims.

4. The good looks of the carefully-printed cotton flour bag are highly acceptable to dealers and housewife.

5. When emptied, the cotton flour bag is greatly appreciated for its salvage value by the housewife who uses it in many different ways.

6. The cotton flour bag can be handled easily and with less fear of breakage.

7. Because of convenience in handling and cleanliness, the baker likes to receive his flour in new cotton bags.

8. Railroads consider new cotton bags, properly loaded in the car, excellent shipping containers.

9. Millers use cotton bags because they assure safe deliveries of flour, hence best relationships with customers.

"Advantages of Cotton Bags for Flour," published in Bagology, August 1927.

LINLINE MATERIAL

THIS BAG MAKES TWO FINE TEA TOWELS,
16" X 30", OR FOUR LUNCHEON NAPKINS
VERY STRONG, LONG WEARING CLOTH.

Minnesota makes the *Best* Flour in the World

49 LBS.

H. H. KING FLOUR MILLS COMPANY

ESTABLISHED 1883

Gold Mine

BLEACHED FLOUR
FARIBAULT, MINN.

REG. U. S. PAT. OFF.

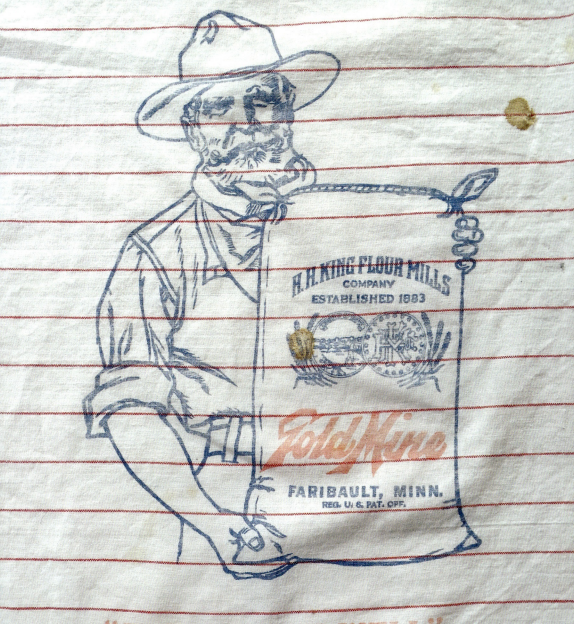

TRADITION AND DEFIANCE

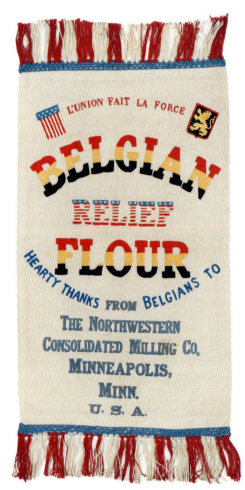

Banner made from a cotton flour sack, 1914–1918.

MINNESOTA HISTORICAL SOCIETY COLLECTIONS

Cotton bags proved valuable in both practical terms and as a political statement during World War I. Filled with flour and other food, they were sent to Belgium to aid the starving population. The empty sacks were subsequently transformed by the Belgians with lace, embroidery and paint. Practicing textile arts on flour sacks provided employment, preserved Belgian traditions and raised money for further relief efforts. It also kept the bags from falling into the hands of the occupying Germans, who wanted to fill them with sand to fortify their trenches and use the cotton to manufacture ammunition.

In 1914, Herbert Hoover (who later served as the 31st President of the United States, from 1929 to 1933) was tapped to head the Commission for Relief in Belgium (CRB). As a highly industrialized country, Belgium traditionally imported 90 percent of its food, and shortages occurred after Germany invaded and occupied the country in 1914. The CRB partnered with numerous mills, many of which sent flour in bags specifically printed for Belgian relief efforts. Between 1914 and 1918, 5.7 million tons of food was sent to Belgium and Northern France, helping to feed an estimated 10 million people.

Textile traditions were strong in Belgium. Women trained in sewing, embroidery and lacemaking at professional schools, and flour sacks provided much-needed fabric for these activities. In addition to bringing in food, the CRB imported thread for lace and exported the finished products, allowing Belgian lacemakers to continue to create and be paid for their world-famous products. Hoover's wife, Lou Henry Hoover, was instrumental in these efforts, and some of the flour sacks were decorated with lace. A number of the elegantly rendered sacks were sold in shops in Paris, London and, for a brief period, New York, raising money for the work of the CRB and increasing awareness of the plight of the Belgians. A select few were

View of a warehouse showing a stockpile of white flour imported from America by the Commission, 1916.

An exhibit of flour sacks, the result of the 1915 newspaper campaign in the United States and Canada. Brussels, United States War Department, National Archives, 1916.

PHOTOGRAPHS COURTESY THE NATIONAL ARCHIVES AND RECORDS ADMINISTRATION, HOOVER PRESIDENTIAL LIBRARY.

Intricately embroidered and embellished flour sacks demonstrate the Belgian's appreciation.

A utilitarian shirt (front and back) made from American Commission Belgian Relief Flour sacks.

HOOVER MUSEUM

given to Herbert and Lou Hoover as gifts of thanks, and some were reportedly displayed in the foyer of the Hoover's home.

Around 400 are in the collection of the Herbert Hoover Presidential Library and Museum in West Branch, Iowa (including some donated over the years). According to curator Marcus Eckhardt, around two-thirds of the museum's sacks incorporate mill logos into their embroidery and embellishment. Some artisans simply stitched over an existing label, while others added images or words that expressed gratitude to America, Hoover and the Commission. In the case of one of the most deftly stitched sacks (right), the ABC mill logo serves as the support for an elaborate rooster standing proudly on a branch, with an American eagle subtly stitched into the rising sun behind. (The image was designed by respected Belgian painter Piet van Engelen, though whether he embroidered it is unknown.) Belgian lace embellishes some sacks, including those used for practical items like tea cozies, carrying bags and pillows. Unadorned sacks also became clothing, many with logos plainly visible. Stains on such a shirt are telling of round-the-clock wear during difficult times. The sacks, it seems, could be elevated to fine art or serve as a resource for everyday life in wartime Belgium.

The wedding quilt of Elizabeth Gould Gardner and James Gardner in the Rising Star pattern, 1850. The backing is of bleached flour sacks.

DEARBORN HISTORICAL MUSEUM

FRUGAL AND FANCY

There is little doubt that resourceful people—some impoverished, some simply frugal—had been reusing fabric from empty bags for as long as it had been available. Nearly every imaginable item was shipped and/or sold in sacks, from mail to laxatives, to sausages, fertilizer and ballots. A tablecloth or bed cover made from salt bags and dating from around 1876 is in the collection of the textile division of the Smithsonian's National Museum of American History, along with a nightgown and corset cover from the later 1800s in the costumes division collection. In the collection of the Dearborn Historical Museum is a wedding quilt, hand-pieced in 1850 in the Rising Star pattern—the quilt top contains more than 1,000 pieces of fabric and the back consists of flour sacks, bleached and sewn together. The Georgia Quilt Project documented quilts from the late 1800s with flour sack backing and the early 1900s with flour sack binding.

Recycling fabric was not only for the working class: thriftiness was a desirable attribute, even for homemakers of financial means. Home sewing was a money-saver, and most middle- and lower-class women had stitching skills. Today, DIY dresses are not necessarily a less expensive alternative to ready-to-wear, but in days gone by, sewing garments and home decor items was cost-effective, especially if the fabric—feed sacks—came "free" with the purchase of goods needed for everyday life.

Cloth bags were favoured as packaging for all sorts of products and purposes.

PAUL PUGSLEY COLLECTION

THE HABHAB BROTHERS' RICE SACK SHIRTS

For most early adopters of sewing with feed sacks—those who turned the white and off-white bags into clothing and household linens—repurposing was not a choice but a necessity. This was the case for Paul Habhab's grandfather, Moussa Habhab, and his brothers Abbas, Yousef and Ali. Some of the first Syrian Muslims to come through Ellis Island, the Habhabs, ages 13, 14, 15 and 16, arrived in the late 1800s and supported themselves by emptying coal from rail cars, using nothing more than shovels and strong backs, for a nickel a ton.

In the 1880s, the brothers moved west to Iowa. "They scattered, working on different farms, and when they got together they would compete to see who was learning the most English," Paul says. "No one could figure out the words one brother was saying, and after a while they realized he wasn't speaking English, it was German, and gave him a hard time. I can see myself doing that, making fun of my brother for learning the wrong language."

What is harder for Paul to grasp is the amount work his relatives undertook to better themselves. From their saved wages, Moussa and Abbas bought a horse and cart, and travelled from farm to farm as peddlers, selling pots and pans, fruits and vegetables. "There was no downtown in those days, so they brought the downtown with them," says Paul, adding that his grandfather was always known for his generosity. "I remember seeing a little ticket book where my grandfather wrote down what he sold and what he was paid—he set up a credit system so that people who were short on money could pay when they were able." The Habhab brothers also sold rice, flour and other items packaged in cotton commodity bags.

Because they, too, needed clothing, Moussa and Abbas made a deal with the women who purchased their goods. "They sold sewing kits with needles and thread, and they'd ask the women who bought them to make them shirts from rice sacks," says Paul. "They'd leave buttons with them and when they came back around they'd pick up the shirts and pay or trade for them. There was a lot of bartering."

By the time Paul was born, his relatives were no longer wearing rice sack shirts and were prominent members of the local community. They married, had families and at various times owned a grocery store, car dealership and gas station. In 1934, the Habhabs were instrumental in establishing the Mother Mosque, the first building designated as a Muslim mosque in the United States. The modest white, wooden structure still stands in Cedar Rapids, and is now listed on the National Register of Historic Places.

Over the years, Moussa and his siblings helped countless family and community members get an education, set up businesses and buy homes. Despite their improved circumstances, Paul says his family always emphasized the importance of frugality. Both his grandmothers had huge gardens, and his mother carries on the tradition, growing food for her family. Paul learned the story of his relatives' rice sack shirts when he asked his father about a pillowcase in their home. "I thought it was a weird print—it was white with blue-ish stripes—and my dad told me it was a rice sack, and the story just came out," he says. "My relatives figured out how to survive, and to get the things they needed through resourceful measures."

Home demonstration clubs were a major proponent of feed sack sewing. Established in the United States in 1917 by the Smith–Lever Act (and today more familiarly known as cooperative extension services), the clubs were part of an effort to educate women on the best methods for caring for their homes and families. They advocated time-saving and money-saving methods based on research done at land-grant universities—American institutions of higher learning that focus on teaching agriculture, engineering and science. (Men attended agricultural demonstration clubs.)

Many small-town papers carried regular columns about home demonstration club activities, and feed sacks featured prominently. In a September 12, 1932, edition of the *Timpson Daily Times* from Texas, a brief article noted, "A dress made of feed sacks at a cost of 54 cents won first place in a contest at the Dilworth Home Demonstration Club in Jim Wells County. Until told differently, every one thought it a linen dress." On July 3, 1936, the *Schulenburg Sticker*, also from Texas, carried a two-column report in their home demonstration club section about Miss Glennie Eilers, winner of the Girls 4-H Bedroom Improvement Contest. In a detailed description of her bedroom ("well ventilated with one door and four windows, the size of glass being 12x30 inches"), Eilers describes varnishing her floor, painting her woodwork, refinishing and repainting every piece of furniture, and hanging wallpaper that complemented her gold and ivory colour scheme. She asked her father to make her a 24- by 26-inch clothes closet, and in front of it she hung a curtain "made out of feed sacks with a border which gives it a clean appearance." She also made a laundry bag out of feed sacks and a mattress cover of "eight good feed sacks," and for her wash stand, dressing table and study table she stitched covers from feed sacks, "making them with a yellow border." Her rocking chair featured "a lean back and pillow to match made out of feed sacks," and she softened another chair with a feed sack pillow with a border. Eilers continues, "I raised a flock of chickens, which covered expenditures of my room and besides I still have a littel [sic] profit." The total cost for the room renovation was $12.08.

HOME CLUB REPORT
OCTOBER 11, 1946

"We were greatly impressed by the hand made garments ranging from infants to adults. They were expertly made—with good lines and style. and what the women have done with feed sacks is beyond belief—one would have to see to realize how cleverly these old bags have been converted into things of use and beauty."

THE WAYNESVILLE MOUNTAINEER

Illustrations from the 1950s booklet Smart Sewing with Cotton Bags.

4-H'ers win sewing honours in a "Sew With Cotton" contest, from the Robesonian, *Lumberton, North Carolina. September 20, 1954.*

"Fashion From Feed Bags" in the Decatur Herald, *Illinois, July 24, 1953.*

On May 7, 1951, the *Waynesville Mountaineer* from North Carolina noted that after giving a presentation on poultry, Mrs. Oliver B. Chason exhibited articles showing the use of feed sacks for making men's shorts and sport shirts. The February 9, 1950, issue of the *Sylva News and Ruralite* from North Carolina quoted a Mrs. Maude Claxton who seemed to feel that teaching wives, mothers and girls to cook, sew and keep house "the right way" was a panacea. She suggested that if homes in her county were happier, "there would be fewer people in our jails, and other places of correction," and that rather than giving clothes to the needy, they should be taught that "they can take old trousers and make a snow suit for their child." She touted feed sacks as ideal dress-making material, and noted that five feed sack dresses had been a part of the previous year's 4-H dress review. "They were lovely dresses, with little or no cost at all," Mrs. Claxton added.

Men's shorts made from a 100-pound Pillsbury's Best feed sack.

Article from the Corvallis Gazette-Times, Oregon, January 23, 1951.

FLOUR SACK UNDERWEAR

FLOUR SACK UNDERWEAR
MORRIS W. JONES

When I was just a maiden fair,
Mama made our underwear;
With many kids and Dad's poor pay,
We had no fancy lingerie.
Monograms and fancy stitches
Did not adorn our Sunday britches;
Pantywaists that stood the test
Had "Gold Medal" on my breast.
No lace or ruffles to enhance;
Just "Pride of Bloomington" on my pants.
One pair of panties beat them all,
For it had a scene I still recall—
Harvesters were gleaning wheat
Right across my little seat.
Rougher than a grizzly bear
Was my flour sack underwear,
Plain, not fancy and two-feet wide,
And tougher than a hippo's hide.
All through depression each Jill and Jack
Wore the sturdy garb of sack.
Waste not, want not, we soon learned
That a penny saved is a penny earned.
There were curtains and tea towels, too,
And that is just to name a few.
But the best beyond compare
Was my flour sack underwear.

BACK TO SACK-CLOTH

The National Cotton council suggests we all go back to the flour-sack underwear to save cotton for defense production.

That would be as bad as going back to home baking -- instead of enjoying better-than-Mother-made PENNINGTON BREAD.

This ad for bread ran in an Ohio newspaper in March 1951.

Flour-Sack Underwear Hailed As Great Aid to Defense Plan

BILOXI, Miss., Jan. 23—(UP)—A return to flour-sack underwear was suggested today as a means of helping conserve cotton for defense production needs.

"Every 100-pound bag salvaged for clothing purposes represents one and one-third yards of cotton cloth released for essential military duty," said N. C. Blackburn of Memphis, Tenn., a member of the national cotton council sales promotion committee.

He addressed cotton industry leaders who are meeting here to discuss ways of meeting the government's goal of 16,000,000 bales this year.

"The homemaker who sews with cotton bags in the face of mobilization restrictions is parallel to planting, growing and processing that amount of cotton for other vital purposes," Blackburn said.

To Encourage Idea
The council will encourage this practice through an educational program in newspapers and magazines and over radio and television.

Frank McCord, the council's market research director, said "under a mobilization program comparable to that of World War II, this country alone would require at least 12,000,000 bales of cotton annually."

Council President Harold A. Young had pointed out earlier that 200 pounds of cotton are required each year for each man in uniform, compared to 20 pounds needed by civilians.

Read P. Dunn, Jr., foreign trade director, told 700 delegates to the 13th annual council meeting that control of the major part of the world's resources would decide the outcome of a third world war.

He said the United States could gain control by mobilizing the industrial output of Japan and western Europe.

Many Uses For Used Cot[ton]

THREE JOLLY PILLOWS

1. When washed, ironed and decorated with embroidery thread, the empty flour bag is easily turned into a beautiful, durable cushion. To remove printing, cover with lard or kerosene and wash the next morning with lukewarm water.

SMART HOME FOR SHOES

2. Any housewife can make a convenient shoe case to hang on the inside of the bedroom door from empty flour bags.

FOR YOUR NEXT PARTY

3. Every hostess should be well supplied with card table covers. They are easily made from empty flour sacks. A little embroidery and binding of the edges is all that is required.

IN the December issue of Bagology we published a r[eview of] Bulletin 582 of the Miller's National Federation enti[tled "] Flour Bags, a Menace to the Flour Trade." The art[icle pointed] out that used or second-hand flour bags were unsanita[ry, a source] of insect infestation, a danger of unfavorable publicity a[nd a source] of damage claims due to breakage. The article continue[d to suggest] that the baker advertise "Empty Flour Bags for Sale" to r[etail] customers.

It is surprising how many uses the second-hand flour [bag can be] put to by the enterprising housewife. On this page we [show] just a few of the more important uses. Certainly, the [use] of second-hand bags in these ways is more economical [than using] them for flour again.

LET KIDDIES PLAY HARD

APRONS FROM

4. All of the little frocks shown here can be made from flour sack cloth. One or two bags cut to your patterns will make a dress or knickers. And they are easily washed and ironed.

5. For durable aprons o[f] the flour sack has an [ ] Flour sack aprons can b[e] week without injury.

To Remove Ink From Flour Bags

1. Smear printed surface thoroughly with lard and let stand over night.　2. Scrub with hot, strong soap suds.　3. Boil
4. Rinse thoroughly.　5. Hang out in the sun to dry and bleach

n Flour Bags

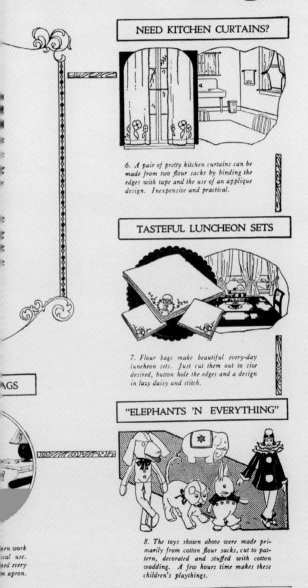

NEED KITCHEN CURTAINS?

6. A pair of pretty kitchen curtains can be made from two flour sacks by binding the edges with tape and the use of an applique design. Inexpensive and practical.

TASTEFUL LUNCHEON SETS

7. Flour bags make beautiful every-day luncheon sets. Just cut them out to size desired, button hole the edges and a design in lazy daisy and stitch.

"ELEPHANTS 'N EVERYTHING"

8. The toys shown above were made primarily from cotton flour sacks, cut to pattern, decorated and stuffed with cotton wadding. A few hours time makes these children's playthings.

"Miss India Hugues, noted Los Angeles beauty, recently registered a most ingenious protest against the high price of bathing suits. She made one from an old sugar sack which cost her only fifteen cents. The suit's charm is said to equal its expense."
Logansport Pharos Tribune, August 9, 1920.

An informative spread from the Chase Bag Company's publication Bagology, *February 1928.*

107

ALBANY DEMOCRAT-HERALD, ALBANY, OREGON, THURSDAY, MARCH 30, 1939. PAGE THREE

Per Friday, Saturday, Monday
Look at this List of VALUES

THE GROCERVETERIA
Free Delivery — C. O. D. Service — Open Evenings till 8 P. M.
PHONE 82

MONTH END SALE
Special Values for Thrifty Shoppers

RINSO Reg. 8c — Large 19c
LUX FLAKES Reg. 9c — Large 21c
LIFEBUOY 3 for 17c
LUX TOILET SOAP 3 for 17c
SPRY SHORTENING 3 lb. can 51c — 1 lb. can 18c
SPRY SHORTENING 6-lb. Tin $1.02

MARY, I FOUND LIPTON'S TEA PEPS ME UP
YES, JANE. JOHN SAYS IT GIVES HIM MORE VITALITY

LIPTON'S TEA FINEST ORANGE PEKOE
1 lb. tin 69c — ½ lb. tin 35c — ¼ lb. tin 20c
LIPTONS GREEN JAPAN TEA — 1 lb. tin 48c — ¼ lb. tin 13c

Willamette Queen HARDWHEAT FAMILY FLOUR	49 lb. Sack	$1.12
VALLEY Blended FLOUR	49 lb. sack	89c
BAKER GIRL hard wheat FLOUR	49 lb. sack	$1.19
KITCHEN QUEEN hard wheat FLOUR	49 lb. sack	$1.22
CROWN Best Patent FLOUR	49 lb. sack	$1.45
DRIFTED SNOW FLOUR	49 lb. sack	$1.45
FISHERS BLEND FLOUR	49 lb. sack	$1.45
GOLD MEDAL Kitchen Tested FLOUR	49 lb. sack	$1.59
CHEESE Whole Milk, Mild, Per pound		14c
CHEESE KRAFT AMERICAN		49c
Handy 2 lb. Loaf BRICK		51c
SWISS		55c
OLEOMARGARINE GEM BRAND, POUND		10c
SALAD OIL Durkee's Fine Food Product	ONE GALLON CAN	95c
COFFEE Todays Brand Vacuum pack tin	per pound	21c
PINEAPPLE Broken Slices Large 2½; Size Tin		2 Cans 29c
BEANS Santiam Brand Cut Stringless		10c
PEACHES H-D Brand Fancy Quality	No. 2½ size can	2 Cans 29c
PEARS Douglas Fancy Bartletts	No. 2½ size can	2 Cans 25c
GRAPEFRUIT H-D No. 2 tall can	11c	DOZEN CANS $1.29
GRAPEFRUIT H-D No. 2 tall can	3 for 25c	DOZEN CANS 95c
SPAGHETTI Franco American or Penthouse	No. 1 tall can	3 For 25c
HALEY'S DICED CHICKEN, 7½-oz. can 33c; 3½-oz. can		17c
HALEY'S BONED CHICKEN, 7-oz. can		25c
HALEY'S CHICKEN SPREAD, 7-oz. can 22c; 3½-oz. can		2 for 25c
RIPE OLIVES, H-D Mammoth, No. 1 tall can		17c
RIPE OLIVES, Dundee large, No. 1 tall can		15c
RIPE OLIVES, Yamhill standard, No. 1 tall can		13c
RIPE OLIVES, Yamhill standard, 4½-oz. can		8c

JOHNSONS FLOOR WAX
YOUR CHOICE
Johnson's Paste Wax, 1 lb. .... 59c — 1-3 pound FREE!
Johnson's Liquid Wax, pint 59c — 1-3 pint FREE!
Johnson's Glo-Coat, pint .. 59c — 1-3 pint FREE!

Fresh FRUITS or Choice VEGETABLES

LETTUCE, Large Crisp heads	2 for 15c	GREEN PEAS, 2 lbs Sweet—Very tender	25c	
CAULIFLOWER Oregon Large white.	Each 14c	Artichokes, 4 for	25c	
ONIONS, Oregon Yellow Denvers	6 lbs. 10c	Carrots, 3 bunches,	13c	
		Green Onions 3 buchs	10c	
ASPARAGUS, 2 lbs Fancy All-Green California	29c	Oranges, 2 doz.	25c	
		Grapefruit, 10 for	25c	
CELERY, lge bunch, Crisp Large Green, Utah	15c	Rhubarb, 2 lbs.	15c	

CRACKERS
Pacific Soda Wafers, slightly Salted, 2-lb. carton 14c
Pacific Brand GRAHAMS 2-lb. carton 18c

BOYS! GIRLS! A GIFT FOR YOU!
Jack Armstrong's "TORPEDO" Flashlight While Supplies Last. With purchase of 2 pkgs.

WHEATIES 21c

ARMOUR'S CANNED MEATS
LUNCH TONGUE, 6-oz. can 21c
ROAST BEEF, 12-oz. can 22c
VEAL LOAF, 7-oz. can 15c
Vienna Sausage, 7-oz. can 9c
CORNED BEEF, 12-oz. can 16c
Cocktail Style Pork SAUSAGE, 5-oz. can 23c
TAMALES, 1 lb. glass 19c
CORNED BEEF HASH, 16-oz. can 15c
SPAGHETTI & MEAT BALLS, 14-oz. can 15c
BEEF EXTRACT, 1 oz. 23c; 2 oz. 39c
DEVILED MEAT, ¼ can, 3 for 10c; ½ cans 6c
SLICED DRIED BEEF IN GLASS JARS,
2-oz. 10c; 2½-oz. 14c; 3½-oz. 17c; 5-oz. 23c

Ginger SNAPS fresh, per pound 11c
FIG BARS, fresh, per pound 9c
CHOCOLATE ECLAIRES, fresh, per pound 20c

CAMPBELL'S PORK & BEANS
16-oz. can — 3 cans 25c

FRUIT JUICES
Buy the Large Sizes and SAVE!
PINEAPPLE JUICE, H-D giant size 46-oz. can 25c
Grapefruit Juice, H-D giant size 46-oz. can 19c
Grapefruit Juice, Garth unsweetned, 46-oz. can 17c
TOMATO JUICE, H-D giant size 46-oz. can 19c
TOMATO JUICE, Armours No. 2 tall can 19c
ORANGE JUICE, Nature Sweet No. 10 (gal) 35c
PINEAPPLE JUICE, H-D No. 10 tin (gal) 49c
GRAPEFRUIT JUICE — Your Choice
TOMATO JUICE — 8-oz.
ORANGE JUICE — Tins — 5c
GRAPEFRUIT JUICE, H-D No 2 can, 3 for 25c
GRAPEFRUIT JUICE, Garth Brand, 3 for 25c
Texas Unsweetened No. 2 can, 3 for 25c
PINEAPPLE JUICE, H-D No. 2 can 9c
TOMATO JUICE, H-D No 2 can 3 cans 25c
PRUNE JUICE, No. 1 tall can 3 cans 25c
APRICOT NECTAR, No. 1 tall can 3 cans 25c

Toiletries REMEDIES — Phone 82
A CONVENIENT SPOT • TO SHOP AND SAVE EVERY DAY

DON'T LET BAD BREATH KEEP YOU APART!
USE PEPSODENT ANTISEPTIC Regularly!
50c SIZE 39c — GIANT 75c SIZE 59c

CLEANSING TISSUES
500 Kleenex 28c
200 Kleenex 2 for 25c
200 Chantilly 10c
200 Handies, close-out 10c
500 Handies, close-out 19c
230 Ponds 2 for 25c
500 Ponds 23c
288 Multicolor Kleenex 25c

Hill's Non-Oily NOSE DROPS 35c No Oily Nose Drops 29c

WE CARRY A COMPLETE STOCK OF ANGELUS LIPSTICKS • ROUGE • POWDER by Louis Philippe IN MATCHING SHADES

BELFAIR Sanitary Napkins 12s 2 pkgs. 25c

We have it! Sunbeam SHAVEMASTER

$1.10 VALUE
ANGELUS LIPSTICK by Louis Philippe
NOW 84c
Thrilling Shades TO MATCH ANY COMPLEXION

Betty K SANITARY BELT WITH SNAPPIES NO MORE PINS — SAFE INVISIBLE SURE
GREATER COMFORT — ABSOLUTE SECURITY
Betty K. Belts Made of Rayon Elastic with Snappies 25c
Betty B. Belts Made of Silk Lastic with Snappies 50c

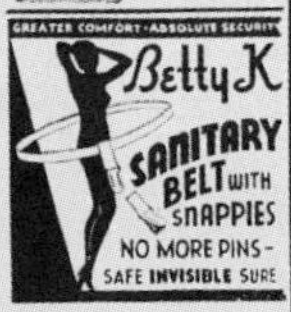

New! TAMPAX 40s 98c
KOTEX 64s $1

MODESS 68c

Modess $1.00
Convenient Terms if Desired
$1.00 Down — 50c Per Week

THE DRY-SHAVER THAT GETS DOWN TO BUSINESS AND DOES A JOB
• Gives you a quick, close, comfort-shave the first time—no skill required — no weeks of patient practice.
• The ONLY electric shaver with a lightning-fast, single cutter that oscillates as an arc inside a comfortable, smooth shaving head.
• The ONLY electric shaver with a shaving head of finest Swedish steel screened to pick up the beard the way it grows.
• The ONLY electric shaver with a powerful, brush-type, self-starting Universal motor, AC-DC........ $15.00

HOPPER'S WHITE YOUTH PACK (CLAY)
55c value, Now 43c

New Line MELLIER TRUE FLORAL ODORS PERFUME
8 Distinctive Varieties
1-oz. 85c — 1½-oz. 50c
¼-oz. 25c — ⅛-oz. 15c

BIG VALUE—ONE PINT GENUINE CLAIR ADAMS ALMOND and BENZOIN LOTION
16 ounces 29c

SUPERIOR PURE U. S. P. SOLUTION Formaldehyde
16 ounces 39c
32 ounces 69c
1 gallon $2.59

25c Kurb Tablets 23c
50c MIDOL 41c
1 Doz. Femeze 25c

Here's a Bargain, Men!
LIFEBUOY SHAVING CREAM
SCHICK INJECTOR RAZOR
8 GENUINE SCHICK BLADES
all 3 for only 59c

Genuine Reg. $1.00 Full Size CHAMOIS 24" x 28" and Regular 50c DUPONT SPONGE BOTH FOR $1.19

Phone 87
CHOICE MEATS at Choice SAVINGS
McDONALD'S MARKET Specials for Fri.-Saturday

PORK STEAK
PORK ROAST lb. 18½c

BACK OR SIDE BACON SUGAR CURED—BY THE PIECE 22c LB.

NEBERGALL'S PURE LARD, 4 lbs. 39c
MUTTON CHOPS---Per Pound 15c
FRESH PACIFIC OYSTERS, Pint 15c

WEARING YOUR BRAND

Although society viewed thriftiness as a virtue for housewives of every economic strata, some still considered feed sack clothing to be a sign of poverty. Early feed sacks were typically sewn into household linens and pieces of clothing that would not be seen. There are apocryphal stories of the young girl who tripped while wearing feed sack underwear to reveal the words "Southern Best!" emblazoned across her rear end, and the wife who made boxers for her husband from a flour sack and left the words "Self Rising" on them.

In North Carolina, people remember seeing clotheslines strung with underwear printed with four Xs, remnants of a logo printed on the sacks used to make the underwear, and a 1946 *Time* magazine article quoted a Pillsbury Flour Company manager as saying, "They used to say that when the wind blew across the South, you could see our trade name on all the girls' underpants."

Leaving trademarks on undergarments was not done for humorous effect—removing the printing was arduous, and included washing and soaking the bags in various concoctions of Fels-Naptha soap, lard, kerosene and bleach. In her book *Mama Learned Us to Work*, Lu Ann Jones recounts a several-day process that started with soaking the bags overnight in Octagon soap and Red Devil lye mixed with warm water. The next day the bags were rinsed, rubbed on a washboard, boiled, soaked in Clorox, rinsed again, then dried and ironed. She notes that the process even entered into southern folklore: some believed that ink detached more easily during a full moon.

Page three of the Thursday, March 30, 1939, edition of the Albany Democrat-Herald (Albany, Oregon) shows the prices of common purchases. Note the 49-pound sack of Willamette Queen Family Flour.

Colgate's Octagon was a laundry bar soap favoured by some women for removing ink from empty feed sacks.

First-- *prepare your cotton bags*

HOW TO OPEN

Cotton bags are sewed with a chain stitch which usually starts in the lower corner near the fold. By cutting the chain close to the bag and taking hold of the ends of the upper and lower threads at this point and pulling both, the stitching can be ripped in a jiffy. Because the thread is of fine cotton, it can be tinted along with the bag fabric should you prefer to work in color.

HOW TO REMOVE PRINTING

Removing the printing on cotton bags has been so greatly simplified by use of new wash-out inks that it is now no trouble at all. Simply soak the bag in warm soapy water. The new inks need little coaxing to do a quick disappearing act. Brand names are sometimes printed on paper labels glued to the bags. Only a few seconds after dipping the bag in water, you can easily strip the label from the sack. Inks and labels should be removed before tinting or cutting.

DYES AND STENCILS

Some of your bags will come in gay and colorful print designs. Your plain bags can take on any enchanting color you wish, merely by using a good all-purpose dye and following the directions carefully. It's fun to plan two-tone effects with your patterns, by dyeing your fabric in contrasting colors. Lots of fun, too, are the designs you can make and stencil on to the finished article. (Be sure you use fabric paints!) This makes a wonderful combination project for the art and sewing classes at school.

Thrifty Thrills with Cotton Bags *offers advice on how to prepare your bag for sewing.*

DEAR HELOISE
OCTOBER 29, 1963
EUREKA HUMBOLDT STANDARD

I remove the printing from my rice, flour and other cotton bags by soaking them in kerosene for several hours.

The bags may be either soaked in kerosene or just dashed with a bit of it, rolled into a ball and left for two or three hours.

I then wash them thoroughly by hand in detergent suds and hot water.

Later, I throw them in my washing machine and run them through the hot water cycle to which I have added a dash of ammonia.

I never hang the bags on the clothesline. I lay them out flat on the grass and let the sun dry them. Sometimes I take the garden hose and sprinkle them. Then I let them dry again.

I wash the bags again in suds to which a little bit of ammonia has been added and lay them back on the grass. All the print comes off my cotton bags this way.

–Sofia

Illustration from a Chase Bag Company ad in Bagology, *August 1947 demonstrating that the company uses washout inks.*

Despite these efforts, logos and wording on fabric persisted. Researcher Ruth Rhoades describes one quilter, Bessie Archer, who used fertilizer sacks on a quilt top and back. Though she dyed the quilt maroon to hide the printing, the pink lettering remained. A master quilter, Archer considered the quilt a failure. (The irony is that the visible graphics make the quilt all the more desirable to today's historians and collectors.) Researcher Pat L. Nickols notes that rather than struggling with logo removal, some women incorporated the sacks' images of flowers, farm scenes and animals as an integral part of their quilt tops. To make sure the images would not fade, they would set them by soaking the sacks overnight in salt and vinegar. Some women then cut and used blocks from the sacks just as they were, while others embellished them with embroidery.

In 1939, the *Salem News* from Ohio ran a story about a campaign by housewives, particularly in rural America and in the South, to make logos easier to erase. Many companies already printed instructions for ink removal right on the bags, but Mrs. Beatrice Humphries, a 43 year old from Judsonia, Arkansas, wanted them to take it a step further, suggesting they use aluminum tags to identify their products. Humphries, who said she had used bags all her life for "every imaginable purpose," noted that even when women followed instructions for removing logos, inks were tough to eliminate: "If … we could only get sugar, salt, or bean sacks that wouldn't necessitate rubbing blisters on our knuckles and holes in the goods on washboards trying unsuccessfully to get out the paint, we would be sold on those products until the end of time." Stories like this helped manufacturers better grasp the extent of women's reuse of feed sacks, and they increased their efforts to make their bags appealing. In 1940, Bemis claimed to have conducted 2,132 experiments on ink removal, and advertised their bags thereafter as being printed with Bemis Washout Inks. The Percy Kent Bag Company touted their 100 percent Soap-Soluble Washout Inks.

"All of the bags are stamped so heavily with ink and paint that it is next to impossible to remove the lettering," complains Mrs. Beatrice Humphries in the Bakersfield Californian, *Saturday, May 6, 1939.*

Women Protest Indelible Trademarks on Flour Bags

MRS. Beatrice Humphries, of Route 1, Judsonia, Ark., believes she speaks for a large percentage of housewives when she says that if flour mills, fertilizer companies and salt and sugar packers would use cotton bags which would lose the ugly patterns in one or two washings, a new era of prosperity would begin for them. She points out that housewives like herself would buy only products packed in bags marked with unindelible inks or stamped with washable paints. Better than that, she advocates no markings on the cloth at all. "Let them use aluminum tags," she suggests.

Mrs. Humphries, a typical Arkansas farm wife, is only one of many southern ruralists who demands her commodities packed in cotton bags. One reason for that is to aid in consumption of cotton. But the main reason is to obtain material for clothing.

"But," she says, "all of the bags are stamped so heavily with ink and paint that it is next to impossible to remove the lettering."

Rather than removing the illustrations on this Blatchford's Chick Mash bag, Henry used them to practise his embroidery skills in 1924.

COTTON BAGS AS CONSUMER PACKAGES FOR FARM PRODUCTS

An excerpt from a November 1933 brochure written by R. J. Cheatham and John T. Wigington and published by the United States Department of Agriculture. "Cotton bags as consumer packages for farm products" was number 173 in their Miscellaneous Publication series.

Cotton Bags Found Satisfactory

The use of various types of consumer packages for marketing farm products has shown that cotton bags are one of the most satisfactory containers. Cotton bags make attractive packages; they supply a suitable surface for brand names and make possible effective advertising; they are durable and little affected by moisture; they represent minimum tare weight; and they have a high salvage value.

Producers, shippers, wholesalers, and retailers recognize the first three of these advantages of cotton bags. They realize that offering farm products for sale in branded packages assists every factor in the marketing chain from producer to consumer.

* * *

Wholesale and retail distributors, and particularly retail grocers, appreciate the improved appearance when products are displayed in small cotton bags. Other important advantages are convenience in storing and handling, saving of time otherwise required for packaging, savings of clerks' time in making sales, and a tendency toward increased sales in some instances through increasing the unit of sale.

The small cotton bag appeals to the housewife, an ultimate consumer, because the trade mark or brand name enables her to identify products with which she has had previous experience. This tends to insure [sic] a continuous market if the potatoes, onions, and other farm products are of high grade. The housewife appreciates the convenience of the consumer-size bag for storing these products in her own home, and she finds that many useful household articles can be made from the cloth in these bags.

Colors Add Attraction

Considerable thought was given to the matter of suitable colors for consumer packages made from cotton fabrics. It was found that bags made from gray goods often become soiled from handling, and by the time they reach the consumer have lost much of their attractiveness. This difficulty was overcome by dyeing the bag fabric.

Sacks of flour in the window of the Golden Rule Store in Mebane, North Carolina. Photo by Dorothea Lange, July 1939.
LIBRARY OF CONGRESS

The interior of a general store in Moundville, Alabama. Photograph taken by Walker Evans, 1936.
LIBRARY OF CONGRESS

A woman walks into a grocery store in Salem, Illinois. Photo by Arthur Rothstein, February, 1940.

LIBRARY OF CONGRESS

STALEY or PENICK **23¢**
SYRUP
White or Golden

THRIFT VALUE!
ROLLED OATS
LARGE 2½ Lb. BOX **10¢**

FANCY MACARONI
SPAGHETTI • ELBOWS • SHELLS

LB. BOX
19¢

STOCK UP!
KROGER'S COUNTRY CLUB
GRAPEFRUIT JUICE
LARGE 46 OZ. CAN **19¢**
Packer's Label Large 46 oz. Can 15¢

Fresh Country
EGGS 19¢ doz.

10 LB.
No. 2
ONIONS
15¢

WESCO 16% DAIRY FEED
SCRATCH FEED
ANALYSIS
DISTRIBUTED BY
WESCO FOODS CO.
CINCINNATI, OHIO

Apples 6 lb. 25¢

Percy Kent Bag Company highlights their needlework imprints as a selling feature to the milling industry.

ADDING VALUE FOR WOMEN

In the competitive cotton bag market it was not enough to simply make inks easy to remove. Bag manufacturers made their bags more valuable by printing "extras" on the reverse side of bags—cut-and-sew dolls, embroidery patterns, quilt patterns, pillowcases and instructions for turning feed sacks into roller towels—a continuous loop of fabric that hung on a wooden dowel. Embroidery patterns stamped onto the bags for days-of-the-week dish towels meant that wives who had stitched Monday and Tuesday towels would encourage their husbands to go back to the feed store to get sacks for the other five days. Sea Island Sugar offered a series of dolls, including Hula the Sea Island dancer, Uncle Sam, Franz the Tyrolean boy, Little Miss Muffet and Abdul the Arabian boy. The Fulton Bag Company devised an "apronbag" in 1937, complete with seam tape sewn inside—the homemaker had only to turn the bag inside out, wash it and cut as per the instructions to create a useful apron.

As early as the 1920s, plain paper labels were attached to sacks with water-soluble glues, then printed with logos and instructions. Other labels encircled the bag and were sewn into its side seams. These were warmly received by busy housewives, who appreciated that they could be removed by soaking them in water, saving time and eliminating the use of harsh detergents and the unpredictable results of the earlier logo-removal process. Manufacturers liked them, too, because their company's name could be displayed on all sides of the bag. In 1944, Bemis patented the Bemis Band-Label Bag, guaranteed to come off in one piece, thus preventing small bits of paper from clogging the household drain.

These sugar sacks offer "needle work for nimble fingers," embroidery patterns featuring various household tasks, from laundry to mopping.

Sea Island Sugar sacks front (with logo) and back with an Arabian boy doll. See page 140 for more sugar sack dolls.

This bag is a ready-made apron—just rip the bag seam, wash and wear! Or, collect a pair and make them into curtains.

The Percy Kent Bag Company filed the patent for this convertible bag and apron design in 1948 and it was granted in 1950.

BEMIS BAND-LABEL PATENT

PUBLICATION DATE
March 2, 1948

FILING DATE
February 21, 1945

INVENTORS
Charles V. Brady,
August F. Ottinger

"This invention relates in general to bags labeled at high speeds; the provision of a labeled bag or the like in which the label is presented as a continuous band around the bag or the like, said band in cases requiring it being reinforced; the provision of a labeled article of this class from which the label may readily be removed and the bag or the like preserved for subsequent use or reuse." Fig. 6 shows a stack of bags, illustrating the better label presentation made by means of the invention.

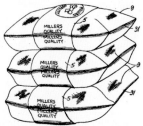

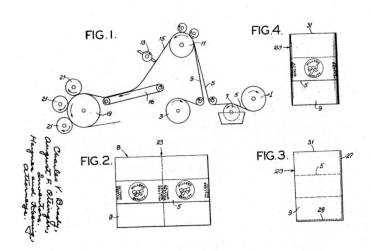

Claims:

1. A labeled bag comprising a rectangular blank of fabric material folded upon itself so that its marginal edges are adjacent each other, thereby forming front and back bag walls. In said seam, the width of the label strip being narrower than the dimension of the bag walls crosswise of the strip.

2. A labeled bag comprising a rectangular blank of fabric material folded upon itself so that its marginal edges are adjacent each other, thereby forming front and back bag walls, the adjacent side edges of the blank opposite the fold and the adjacent bottom edges of the blank being in, turned and stitched together to form inturned side and bottom seams, a continuous printed paper label strip peripherally completely encircling the outside of the folded blank from one of its adjacent side edges to the other with its mid portion traversing the fold, said strip being secured throughout its area to the blank by a water soluble adhesive thus permitting the strip ultimately to be removed from the blank, the ends of the strip being inturned with the side edges of the blank and caught in the side seam, the width of the strip being narrower than the dimension of the bag walls crosswise of the strip.

3. A labeled bag according to claim 1, wherein one longitudinal edge of said label strip includes a reinforcement which is also inturned with the seam at said edges and caught in said seam.

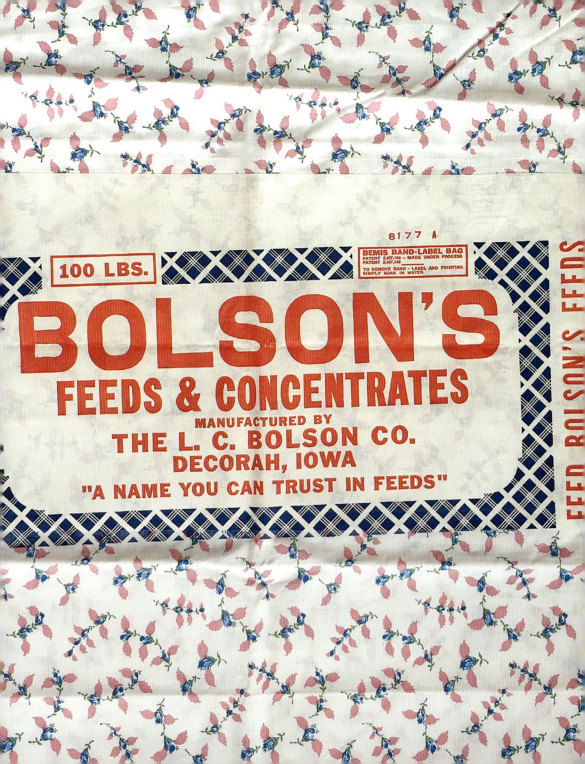

Among the simplest incentives to add value were embroidery guidelines printed on sacks. Once stitched and washed, the printing faded.

The example at left is a sack offering an apron. In this case, the maker has appliquéd the floral motifs.

This sack offers a baby's bib and slippers to be cut, sewn and embroidered.

According to the publication *Feedstuffs*, a trade magazine founded in 1929 that covers animal agriculture, the success of these types of endeavours showed that there was value in "attracting feminine attention to get the masculine dollar." In their advertising, manufacturers reassured their customers—the flour mills and sugar, seed and feed companies—that bags were nearly as important to consumers as the contents of those bags. A Percy Kent Bag Company advertisement noted, "It's the 'extra' that often determines her preference for your product," and a 1933 United States Department of Agriculture brochure assured producers of potatoes, oranges and other agricultural products that "cotton bags make attractive packages; they supply a suitable surface for brand names and make possible effective advertising; they are durable and little affected by moisture; they represent minimum tare weight; and they have a high salvage value."

These marketing efforts were especially important at a time when changes in technology and farm practices were resulting in a waning need for cotton bags. As farms grew in size, obtaining feed and seed in bulk became more common. Paper bags (produced by Bemis as early as 1913), duplex bags (which had mesh on the front and were solid on the back) and cardboard cartons had come to be seen as more sanitary and convenient, and gained a foothold with rural and urban customers. Families also consumed less flour, which was one of the most common contents of cloth bags. These changes were already on the horizon by 1925, when manufacturers united under the Textile Bag Manufacturers Association to protect their interests.

STITCH FIRST THEN TO REMOVE INK EASILY SCRUB THOROUGHLY WITH WARM WATER AND NAPTHA SOAP. DO NOT SOAK OR PUT IN EXTREMELY HOT WATER.

Seasonal baking was extra sweet when bags featured toys to cut out and sew for children, and decorations such as Christmas stockings.

The sack above has a full-colour story book printed on the back.

The Imperial Flour bag manufactured by Percy Kent includes a dapper-looking Sunday pup on the back.

SHIP'G BAG STD BY
PERCY KENT BAG CO.

Tea towel
NEEDLEWORK PRINT
ON THE BACK
Cambric Cloth

Cotton Bags
MAKE THESE PRACTICAL THINGS

LUNCHEON SETS	APRONS	HOUSE DRESSES
PAJAMAS	STUFFED ANIMALS	PILLOWS
PILLOW CASES	POT CLOTHS	TRAY CLOTHS
ROMPERS	HANDKERCHIEFS	MUFFIN COVERS
SUN SUITS	DISH TOWELS	HOOKED RUGS
SMOCKS	LAUNDRY BAGS	BEACH COATS
CURTAINS	CHILDREN'S APRONS	CRIB COVERS
MATTRESS COVERS	BEDSPREADS	BIBS
VANITY TABLE DRAPES	COLLAR AND CUFF SETS	
IRONING BOARD COVERS	CARD TABLE COVERS	
TABLE RUNNERS	NEEDLE WORK DESIGNS	

Guaranteed

48 LBS.

IMPERIAL

FLOUR

WALNUT CREEK MILLING CO.
GREAT BEND, KANSAS

bleached

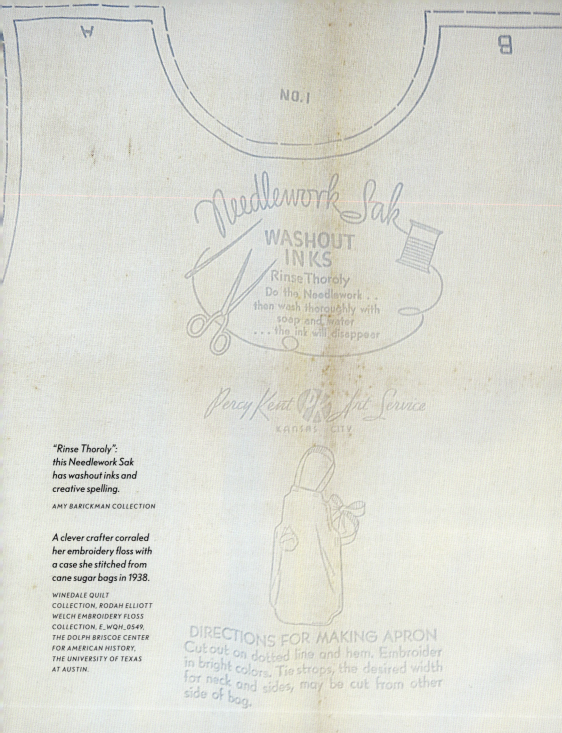

"Rinse Thoroly": this Needlework Sak has washout inks and creative spelling.

AMY BARICKMAN COLLECTION

A clever crafter corraled her embroidery floss with a case she stitched from cane sugar bags in 1938.

WINEDALE QUILT COLLECTION, RODAH ELLIOTT WELCH EMBROIDERY FLOSS COLLECTION, E_WQH_0549, THE DOLPH BRISCOE CENTER FOR AMERICAN HISTORY, THE UNIVERSITY OF TEXAS AT AUSTIN.

The Betty Bemis doll came printed with undergarments and shoes. There was enough room on the bag to also include two dresses.

The gentlemen are Ceresota Flour advertising dolls (dates unknown).

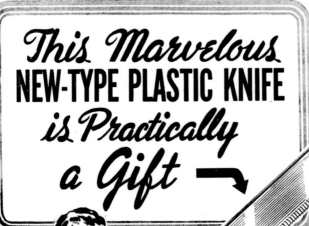
This Marvelous NEW-TYPE PLASTIC KNIFE *is Practically a Gift*

Never Retailed for Less Than 69¢

Yours for only 25¢ and Two Coupons *from* GOOCH'S BEST WORLD RECORD FLOUR

Thousands of good cooks are switching to the world record holding Gooch's Best Flour, that so many champions are using in their baking. This fine, uniformly even-textured flour has established a *world record* no other flour can claim. White bread baked with Gooch's Best Flour has been judged the best and awarded the first prize 28 times at the Nebraska State Fair. Seven times as many first as all other flours combined.

The reason for this remarkable record is in the milling and selection of wheat used in the manufacture of this prize winning flour. GOOCH'S BEST is blended especially for home size recipes—blended to make baking easier and results more certain. GOOCH'S BEST All Purpose FLOUR is milled from choice Nebraska, Kansas and Colorado wheat containing just the kind of good mellow gluten that makes mixing easy and holds the dough just right for really fine breads. You don't have to catch the dough at the exact moment for baking to get perfect results when you use GOOCH'S BEST. That's why it's now famous as the Flour of Champions—the World Record Holder. Do as so many good cooks are doing—say "Gooch's Best All Purpose Flour" when ordering from your grocer.

Unbreakable, Stainless Knife Amazes Housewives!

Just imagine!—a knife that sections grapefruit, slices and pares fruits and vegetables, cuts pies and cakes, even slices bread evenly ... yet never needs sharpening.

Unlike the ordinary steel knife, it leaves no after taste to fruits, vegetables and pastries ... It don't chip or break ... It's stainless and easy to clean and it's all in one piece so you'll never have a loose handled knife. Best of all, children can use it without cutting their fingers or hands.

All these features are in the new crystal-clear plastic knife that every housewife can have for only 25c and two coupons from GOOCH'S BEST FLOUR. Never before has this truly remarkable knife been offered anywhere for less than 69c. Small wonder, though, for it's a knife that should last a lifetime and one that takes the place of several knives.

It's a revelation in modern cutlery and now you can get one so practically as a gift. They won't last long, so hurry, send for yours right away!

QUANTITY LIMITED GET YOUR KNIFE NOW!

PACKED IN COLORFUL USABLE DRESS PRINT BAGS

These Gooch Dress Print sacks are really quality materials ... closely woven and of smooth texture. They are color-fast. You'll find them practical for making literally hundreds of useful and nice looking things which you'll be proud to own—things such as dresses, children's clothes, drapes, quilts, curtains, luncheon sets and many others. You'll be thrilled with this extra saving and value you get when you order GOOCH'S BEST ALL PURPOSE FLOUR.

GOOCH MILL & ELEVATOR CO.
LINCOLN, NEBRASKA

FIRST AGAIN—
White Bread Baked By Mrs. Milton Gates
Made With
GOOCH'S BEST FLOUR
Won First At The
1941 NEBRASKA STATE FAIR
28 Times out of 32 GOOCH'S BEST Flour has helped some Nebraska house win this award.

WINS CAKE PRIZE TOO
Yes, some of the most difficult cake premiums to win such as Angel Food, Devil's Food, Coconut Layer and many others were won by skilled home bakers who used GOOCH'S BEST Flour in their baking.

Let the success of these prize winning home bakers guide you in your choice of the flour to use in all your baking.

INCENTIVES

FLOUR, FEED AND SOMETHING MORE

In addition to printing embroidery patterns, dolls and dress fabrics on cotton bags, feed sack companies lured buyers with additional premiums. Some items like decorative needle books and thimbles were given away at mills and grocery stores, while other items were included with the contents of the sacks. Recipes and recipe books were available from Occidental Flour and Mother's Best Flour, and Mother's Best also included silverware coupons in every sack. One company offered a nine-inch zipper with each sack.

Collector Gloria Hall learned that some manufacturers packed glass pitchers in quart and half-quart sizes for pouring pancake batter along with their flour. Gooch's Best Flour offered a plastic knife for 25 cents plus two of the coupons found in their bags of flour—an advertisement noted that "this marvelous, new-type plastic knife" could not be purchased anywhere else for less than 69 cents. Ballard's Obelisk Flour of Louisville, Kentucky, offered tinware that ranged from a teaspoon, pepper box, grater or biscuit cutter in their 12-pound sacks of flour, to a milk pan, quart bucket, funnel or match safe in their 48-pound sacks. The ad noted: "This will not add to the price in any way, but is merely given to our customers to get the flour well introduced."

★ For the LADIES

100 Dress Print Bags to be given away. 4 Winners will be drawn every hour from 9 A.M. until 5 P.M. for a total of 36 winners. Each to select 3 bags of the same pattern.

Others drew a crowd simply by giving away empty feed sacks. A May 3, 1955, ad in the *Hope Star* from Arkansas for the grand opening of the Darco Farm Store promised that along with an appearance by "Aunt Lucy and her dog Weegee from KARK-TV," they would be giving away 100 dress print bags. Four lucky winners' names would be drawn hourly, from 9 a.m. to 5 p.m., and each of the 36 winners would receive three bags of the same pattern. (The event's grand prize was a drawing for 500 pounds of free feed, and everyone who attended received "FREE Ice Cold! Dr. Pepper.")

Gooch's Best gave their customers glass pitchers for pouring pancake batter as well as plastic knives.

Ballard & Ballard's advertisement of the various free incentives inside their barrels and sacks of flour.

A promotional thimble for the I-H Grist and Flour Mill in Jamesport Township, Missouri, from around 1920.

Needlebooks advertised flour and farm feeds while providing useful needles for home sewing.

This 25-pound bag of PurAsnow Enriched Self-Rising Flour was packed in a sack with extra fabric (note the chain stitching sealing the bottom and side of the bag). In addition to the 15 cents off, the bag provided a home sewist additional yardage.

THE GREAT DEPRESSION

Oklahoma squatter's family with her children at the entrance to a squatter's shelter. Photograph by Dorothea Lange, 1935.

Children of a family living on a grazing land project. Oneida County, Idaho. Photograph by Arthur Rothstein, May 1936.

Opposite: A Dorothea Lange portrait of the wife of a migratory labourer with three children. Near Childress, Texas, June 1938.

LIBRARY OF CONGRESS

The Great Depression (1929 to 1939) helped counteract the waning interest in cotton bags. In the United States, between 13 and 15 million people lost their jobs and nearly half the banks failed. Dot Bloom, who graduated in 1937 from high school in Exeter, New Hampshire, remembers not only reusing feed sacks, but taking apart outgrown clothing, particularly wool skirts, and using the less worn, reverse side to construct a garment for a younger, smaller sibling. "We were proud of doing that. We'd say 'Look, I got a sleeve out of this little piece,'" she says. "We were a middle-class household, but no one had any money and no one looked down on anyone else. Everyone used anything we could get hold of, and you were lucky if you had access to it."

In "Thank the Lord for Feed Sacks," an article in the September 11, 2002, issue of the *Rogersville Review* from Tennessee, interviewee Sue B. Wallin said her mother was grateful for feed sacks during the Depression. Without them, she says, "most little younguns would have run around naked."

Concurrent with nationwide economic devastation was the drought that created the Dust Bowl, which in the United States rendered more than 100 million acres of farmland arid and unusable. It also forced thousands of rural families in Oklahoma, Texas and other plains and Midwestern states to give up their land, load their belongings in cars and trucks, and head to California in search of work. The "free fabric" of feed sacks was critical for people in these circumstances, and photos taken by Dorothea Lange and others of these migrants, along with photos of southern sharecroppers, show many wearing garments likely stitched from feed sacks.

A mother in California, waiting with her husband and two children to be returned to Oklahoma by the Relief Administration. The family had lost a two-year-old baby during the winter as a result of exposure. Photo by Dorothea Lange, 1937

LIBRARY OF CONGRESS

Farm girl near Morganza, Louisiana, photographed in 1938 by Russell Lee.
LIBRARY OF CONGRESS

Farmers sitting on bags of rice at a mill in Louisiana. Photo by Russell Lee, September 1938.

LIBRARY OF CONGRESS

COTTON BAGS MEAN JOBS

A worker in the Lauren cotton mill, Mississippi, repairing a break in a spool of thread. Photograph by Russell Lee, 1939.
LIBRARY OF CONGRESS

Use Of Cotton Bags Means More Jobs For Workers

A carload of sugar packed in cotton bags provides a grand total 766 hours of labor and also aids the cotton industry, according to statistics compiled by W. W. Overton & Company, Dallas.

In a message to grocers intitled "cotton bags mean jobs," the company points out that one carload of sugar means 16,000 5-lb. bags in 800 containers. These take 9-10 bales of cotton, the product of 4 1-2 acres of ground. Raising the cotton takes 414 hours of farm labor and making it into 3578 yards of goods takes 220 hours of cotton mill labor. Making the cloth into bags takes 132 hours of bag factory labor. All this reached a total of 766 hours of labor.

Officials of the Overton concern declare the American public consumes approximately 162,000 carloads of sugar per year, of which some 120,000 carloads are for home consumption. This requires a year's work and income for 46,000 people and consumption of 228,000 bales of cotton, officials said. In all the statement shows some 12,000,000 people depend for their livelihood on the continued use of cotton. These people are the grocer's customers. Overton officials said the housewife customer appreciates the reuse value of cotton bags.

"INSIST ON COTTON BAGS"

W. T. McGehee of the Goyer company of Greenville passes on to us a broadside, printed on cotton sugar bag material, with the caption "Insist on Cotton Bags." It comes from the Cotton Textile Institute, and is as effective an editorial for use of cotton products—and especially for sugar bags—by consumers as could be written. We quote it in part:

Who'll Get the Money?

When money flows into your community it brings good times. When it goes somewhere else it leaves unemployment, hunger and distress for you and your neighbors. Think of this when you buy staples such as sugar, flour, meal and salt. Make sure YOU buy these COTTON bags. Every time a carload of sugar, for example is sold in COTTON bags it pays a day's wages to 83 people who depend on cotton for a living. Unless you indsist on COTTON bags these wages go elsewhere and your neighbors and you must suffer.

The plea of the Institute is especially timely for there is a marked turn to paper bags by many packers of sugar, flour, salt and other commodities. The Institute's message continues with these figures:

Eighty thousand pounds of Sugar (1 carload) if packed all in cotton bags and containers will contain 16,000 "five pound Cotton Pockets, and 800 Cotton containers. This equals the production of

---

Merchandise your Products in Cotton Bags and Wrappings

Cotton Week
1 Bale of Cotton
— — GIVES — —

218 Hours of Work To A Sharecropper

Making this cotton into 2,270 yards of cloth gives

126 Hours of Work to a Textile-mill Operative

Making this cloth into 13,600 5-lb. sugar bags gives

106 Hours of Work to a Bag-factory Employee

COTTON BAGS MEAN JOBS

GROCERS need the business created by these jobs.

Greater and continued use of COTTON in whatever form helps meet one of America's most important economic problems.

1. COTTON gives WAGES to far more people than its substitutes. 2. These WAGES buy grocers' commodities. 3. Lack of these wages hurts buying power in all lines. 4. Using COTTON for bags gives the agricultural field a substantial market and buying power. 5. The combined movement creates more WAGES, more markets, and better times for all industries.

Your housewife customer has a definite use for the COTTON bags you provide. She will appreciate them. See that the men responsible for buying your sugar, flour, salt, meal, and kindred items realize this condition. Have them start now to see that these commodities are purchased only in COTTON bags for your trade.

Make Every Week - - - National Cotton Week

Corsicana Cotton Mills

M. E. WOODROW, Vice-President and General Manager   E. E. SHEEHEY, Secretary and Treasurer

---

NATIONAL COTTON WEEK MAY 22-27

THE FIBRE OF AMERICAN PROSPERITY

Use More Cotton Bags

While the increased use of cotton in all fields will be urged during National Cotton Week, May 30 to June 4, perhaps more attention will be given to stimulating the use of cotton bags for food and other projects, than to any other in particular, since it is one in which citizens in every walk of life can well be interested.

It is, perhaps, asking too much of American women to wear cotton hosiery instead of silk, even in a protest against the activities of the people from whom the bulk of our silk supply comes. Attention, however, should be given to the more general use of cotton fabrics for wear by men, women and children, and for a wider use of cotton fabrics in the home, instead of rayon and other materials which are in competition with cotton.

There are many arguments for the use of cotton bags as food containers. They make an attractive package, are stronger than paper, are durable, permit "natural breathing" of contents and prevent caking, especially important in such products as salt and sugar, and are valuable after they have served the primary purpose. Many housewives realize this and make use of cotton bags, for one thing or another, when they have been emptied of their original contents.

Speaking of this matter of boosting the use of cotton bags for food products, and calling attention especially to the value of this position by those whose livelihood is dependent on the cotton industry, The Journal of Commerce and Commercial says:

It has been estimated that an 80,000-pound carload of raw sugar if packaged in cotton bags would provide a day's work for 83 persons. The same carload of sugar if packed for distribution in paper bags would give a day's work to but eight persons.

Insistence by those whose livelihood is dependent upon

SAVE THE STRING

The String, Too, Is Worth Saving

One reader has made a crocheted tea cloth out of the string she saves when she rips the cotton bags that come into her home. The bag is sewn with a chain stitch that

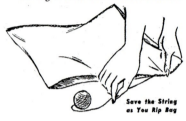

Save the String as You Rip Bag

usually starts in the lower corner near the fold. By cutting the chain close to the bag and taking hold of the ends of the upper and lower threads at this point and pulling both, the stitching is ripped in a few seconds.

Advice on saving string, from A Bag of Tricks for Home Sewing.

In the era of "waste not, want not," women saved the heavy cotton thread used to stitch bags shut along with the fabric. A 25-pound bag could yield 240 inches of string. Once manufacturers realized women were reusing feed sacks, they began printing instructions for accessing the bag's contents directly on the sack. Pointing hand motifs or arrows would indicate: "To open easily cut off string and unravel from this end." The instructions printed on a sack of the C&H company's Pure Cane Berry Granulated Quik-dissolving Sugar included colour-coded strings: "To open cut ALL strings Pull BLUE string." This unique feature, for which the manufacturer applied for a patent, made bags even easier to open.

With an eye toward using up every last bit, feed sack string was put to myriad uses. Phyllis Rosenwinkel recalls that there were always balls of feed sack string around her family's Iowa farm, where it was used to wrap presents, to tie plants to poles and even to sew patches on her father's work pants.

Early feed sack string was cotton and took dyes well, so women often dyed it to match a coloured fabric they were sewing with. (Later feed sacks were sewn with rayon string.)

Dot Bloom, who grew up during the Great Depression, remembers that her mother never sewed pieced quilt tops to batting and backing, but instead tied them with the thick thread she saved from "grain bags" (the name that Dot says sacks were called in New England). Creative homemakers used the string to crochet table runners and doilies, and there are accounts of socks knitted from the conserved threads (although what kept them from falling down around the wearer's ankles remains a mystery).

Whether used for decorative or utilitarian purposes, feed sacks brought out women's creativity—as well as a bit of playfulness. Loretta Smith Steele of West Virginia remembers that at Christmastime she and her siblings each got the gift of a coconut. Loretta's father created

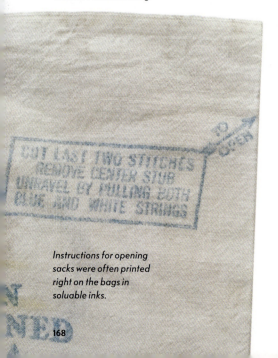

Instructions for opening sacks were often printed right on the bags in soluable inks.

How to Make Hairpin Lace

Step 1—Make a loop at end of ball thread (Fig. 1).
Step 2—Insert hook in loop and wind ball thread around right prong of staple (Fig. 1).
Step 3—Thread over hook and draw through loop, keeping loop at center (Fig. 1).
Step 4—Raise hook to a vertical position and turn staple to the left (Fig. 2).
Step 5—Thread over hook and draw through loop on hook (Fig. 3).
Step 6—Insert hook in loop of left prong (Fig. 4).
Step 7—Thread over hook and draw loop through (2 loops on hook), thread over and draw through 2 loops.
Step 8—Repeat Steps 4 to 7 inclusive until staple is filled.
Step 9—Remove all but the last 4 loops from staple and continue as before for desired length.

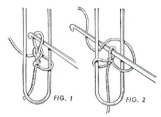
FIG. 1   FIG. 2

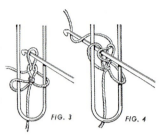
FIG. 3   FIG. 4

a mask by drilling holes in the coconut husk for eyes and a mouth, and then threaded feed sack string through the mouth. The result was what he called a "monkey" string-holder.

The "women's section" of local newspapers sometimes highlighted ways to use feed sack string. These sections first appeared in the 1890s, and included tips on the "womanly arts" of cooking, furnishing a home and sewing—the *Kansas City Star*, for example, printed 1,068 quilt patterns from 1928 to 1961, and undoubtedly many were used as the basis for feed sack quilts. The National Cotton Council seemed especially adept at placing articles in women's sections about the usefulness of feed sacks. In 1947, the *Madison County Times* of Chittenango, New York, highlighted the reuse of feed sack string for making hairpin lace, a technique that employs a crochet hook and simple lace loom (two metal rods or wires) for making shawls, baby blankets and decorative trims.

> **Feed Sack Handicraft**
> Hairpin lace can be made from thread unraveled from feed sacks and used in the ornamenting of handicraft articles which are made from feed sacks, according to home industries specialists.

When selling their sacks to industrial packagers like flour mills, some bag companies also sold the string used to stitch the filled sacks shut.

At right, a simple doily made from sack string.

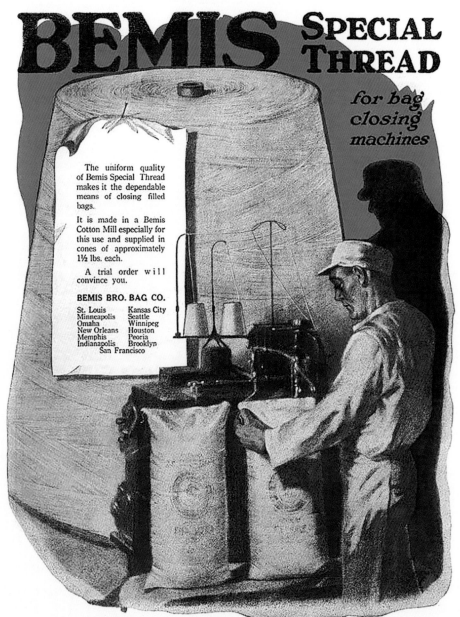

SAVING MONEY With Cotton Bags

A trim and tailored utility dress can be made from laundered Cotton Flour, Sugar, Feed or Meal Bags. The shirtwaist frock illustrated is pretty enough to dress up in any afternoon and yet comfortable enough to be worn doing the heavier morning chores. No trimming is necessary — but rick rack or colored bias tape will add color and effectiveness. With a little variation the same pattern can be used for a number of dresses. The material in Cotton Bags can be dyed or tinted to any shade you desire—and a little starch makes it crisp and smart.

Other Cotton Bag Sewing ideas are illustrated in a free booklet. Write to National Cotton Council, Box 18, Memphis, or Textile Bags, 100 N. LaSalle St., Chicago.

SAVING MONEY With Cotton Bags

The clever housewife these days knows how to be her own attractive self at all times, even when doing her general cleaning. A large coverall apron made from cotton bags is comfortable to work in. The apron pictured is plain and very practical and easy to make. Also illustrated is a headband to keep milady's hair clean and in place. A cover for your broom will make it possible for you to reach over door and window frames and mouldings to get dust. A mitten sewed to your dust cloth, with the wrist left unsewed, forming a pocket in which to slip your hand, is a marvelous protection for your hands and nails. All made from cotton bags, available through your own purchases, or from your nearest baker or department store.

Other Cotton Bag Sewing ideas are illustrated in a free booklet. Send to National Cotton Council, Box 18, Memphis, or Textile Bags, 100 N. LaSalle St., Chicago.

of 1939

SAVING MONEY With Cotton Bags

The peasant influence is new and interesting. One of the smartest aprons we've seen this season is in the dirndl manner. This was made from one large Cotton Bag. Use the full width of the Bag but cut a strip from the length of the Bag to be used for the belt and sash ends. The apron is gathered on to the tight belt and several rows of brightly colored tape or rick rack are sewed in rows across the bottom about three inches apart. Or a cross stitch design in bright colors is an effective trim. If you do not have a large Cotton Bag ready for use, one can be purchased for just a few cents from your grocer or baker.

SAVING MONEY With Cotton Bags

Easy to make—and pretty enough to please any little girl from 2 to 8 years old. And it's made from laundered cotton bags! This little dress for play or school is cut in six gores, with tiny puff sleeves. Use just two large cotton bags that originally contained flour or sugar. They are soft and white when laundered. The neck and sleeves and bottom of the skirt are bound with colored bias tape, with an extra row about an inch above the bottom edge, and little colored buttons to match the tape the only extra trimming needed. Only a few cents—and how delighted the small daughter will be! Extra cotton bags can be obtained from your nearest baker or department store.

Other Cotton Bag Sewing ideas are illustrated in a free booklet. Send to Textile Bags, 100 N. LaSalle St., Chicago, or National Cotton Council, Box 18, Memphis.

SAVING MONEY With Cotton Bags

Brighten up your kitchen to help yourself enjoy the many hours spent there. Colorful Mexican designs or the working bears illustrated are worked in simple embroidery stitches. A large Cotton Bag makes a serviceable towel—if you like the smaller size cut the bag in two and hem the unselvaged sides. Cotton Bags make ideal dish towels because they are soft, absorbent and durable, and leave no lint on dishes.

Other Cotton Bag Sewing ideas are illustrated in a free booklet. Write to National Cotton Council, Box 18, Memphis, or Textile Bags, 100 N. LaSalle St., Chciago.

SAVING MONEY With Cotton Bags

A dainty little Pinafore to please a dainty little lady! And Pinafores for tiny tots and big sisters are one of the newest fashions. Make

SAVING MONEY With Cotton Bags

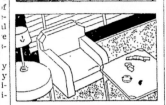

Are you upset about the cigaret burn in the leather top of your new bridge table? Don't worry about it—because a washable cover will hide the scar. One large cotton sugar bag or flour bag will make a cover just the size of the bridge table. A bolt of bias tape can be used to bind the edges and make the ties to hold the cloth secure. Add a little starch when laundering, and your cotton bag bridge table cloths look like linen and gives a smooth playing surface which resists soil.

Many cotton bag sewing ideas are illustrated in a free booklet. Send to National Cotton Council, Box 18, Memphis, or Textile Bags, 100 N. LaSalle St., Chicago.

SAVING MONEY With Cotton Bags

Keep clothes pins in the large pocket of this clothes pin apron while hanging up the laundry. It really makes the task much easier. Make it of one large or two small Cotton Flour, Sugar or Meal Bags. Cotton Bags, when laundered, yield a piece of white, durable material —and the only trimming necessary for this practical apron is a few yards of bias binding tape. The clothes pin design on the pocket is stitched with colored thread to match the binding.

SAVING MONEY With Cotton Bags

Snowy white cotton underwear beneath sheer cotton blouses is a new note this season. Slips and combinations are quickly and easily made at practically no cost. Make them of laundered Cotton Bags, which are soft, white and durable. Keep the lines of your "foundation" smooth and trim. For a gay note, add a flounce of lace or embroidery.

Other Cotton Bag Sewing ideas are illustrated in a free booklet. Write to National Cotton Council, Box 18, Memphis, or Textile Bags, 100 N. LaSalle St., Chicago.

SAVING MONEY With Cotton Bags

Protect your bedding with covers made from Cotton Flour, Sugar, Feed or Meal Bags. The coil springs cover protects the mattress from rust stains caused by the springs. The mattress cover helps prevent the soiling and fading of the mattress. The quilted mattress pad shields the mattress from wear and helps to keep it clean — also makes a softer, smoother bed. Pillow protectors and pillow slips can also be made of Cotton Bags. All these covers can be washed easily and your bed will always be fresh and clean.

Other Cotton Bag Sewing ideas are illustrated in a free booklet. Write to National Cotton Council, Box 18, Memphis, or Textile Bags, 100 N. LaSalle St., Chicago.

SAVING MONEY With Cotton Bags

Summer cottages may be charming. A great variety of articles especially suitable for the summer home can be made from used Cot-

SAVING MONEY With Cotton Bags

Old fashioned Hooked Rugs add charm and color to any room in the house. Make them of Cotton Flour, Sugar, Feed or Meal Bags. Dye some of the Cotton Bags with a fast color dye, bright colors preferred. Leave a few of the Cotton Bags their natural white to mix with the color. Cut the material into ¾" strips, then sew the strips end to end and twist the strips as you crochet. Use a heavy wooden crochet hook, say a size 11. Make the rug any size and shape to suit

Woman's World
Using Cotton Bags for Garments Reduces Fabric Cost to Trifle

By Ertta Haley

THESE days there's a price tag on everything, so when we do come across something free we really stand up and cheer. What is free, you ask? Flour bags, for instance.

Those of you who buy flour or feed in large quantities know that the material in the bags is well worth sewing effort, and they can be made into some of the most attractive garments you've ever seen. I've seen many attractive prints in such bags that make lovely blouses or table linen. And that doesn't even begin to scratch the surface of their possibilities.

Don't scorn the white bags either, not with all the attractive colors that are available in dyes. Two or three of the white bags can be dyed at the same time, and there's at least three or three and a quarter yards of material there for an attractive school dress for the teen-aged daughter.

Most women recognize the value of flour, feed, sugar and salt bags, but too often their eyes are closed to their possibilities for anything except towels. However, experience proves that we can use bag fabrics for not only those things already mentioned but also skirts, draperies, coveralls, lingerie, pillows, etc. Just ask some bright 4-H girl how many things she can name that can be made from cotton bags, and she'll amaze you.

Cotton Bags Make Year-Round Dresses

You don't have to confine wearing apparel out of cotton bags for just the warmer months. After all, cottons are a year-round fabric. Before you begin a dress project, for example, collect enough of the bags so you can cut out an attractive pattern. Then rip these apart and dye them carefully to a shape that will fit you perfectly.

Collect several cotton bags...

Drum Bonnet

A drum bonnet, designed by Sally Victor, is made of smoky white felt with a band of gold braid across the front of the crown to accent the mink scarf that forms a snug wrap around the shoulders.

material, and the results will be well worth the effort.

How to Prepare Cotton Bag Goods

Get the whole family to help you in collecting the bags and make sure you have enough of them before you start on your pet project. As soon as you get a bag, rip it apart and remove the labels. The different type labels respond to different methods of removal. On some bags, they can be removed by soaking in warm soapsuds overnight. Preliminary scrubbing, soaking and boiling will take out others.

Exposure to direct sunlight is a good method of bleaching the bags in many cases.

When labels do not respond to the above outlined treatments, soak the bags overnight in kerosene or turpentine, then wash in soap and water. If any color remains, treat them with a commercial bleach. Always remember which treatment works best with certain kinds of bags so you can cut out an attractive pattern. Then rip these apart and dye them carefully to a shape that will fit you perfectly.

Collect several cotton bags...

Dry the material and press very carefully. It's no fun trying to cut and sew something that looks worn and wrinkled, so prepare the material to look like freshly cut bolt goods.

Some of the choices which you can sew beautifully from cotton bag fabric are these: A tailored but casual dress with long or three-quarter length sleeves with contrasting colored stitching as the main trimming; a full-skirted dress with touches of dainty eyelet or lingerie trimming in the sleeves and at the edge of the skirt; a tailored dress made of two different colors that blend well together. This may be print and plain material combined, or two solid colors combined.

Lay out the pieces of your pattern on the material before you do any cutting whatsoever. The fabric guide of the pattern will not be of too much help, inasmuch as you are using material which will not fit regulation yard lengths. Do not cut until you have fitted everything together and then pinned the pattern to the fabric. Pinking shears are ideal for finishing the edges of this cotton material.

To make pretty dresses.

After the material is cut, sewing is done just as you would on any other dress, no matter what the fabric. Just because you are using cotton bags, however, don't get careless with the sewing. Give it all the care you would if you were making a dress out of ten dollar a yard fore you start on your pet project. As soon as you get a bag, rip it apart and remove the labels. The different type labels respond to different methods of removal. On some bags, they can be removed by soaking in warm soapsuds overnight. Preliminary scrubbing, soaking and boiling will take out others.

Exposure to direct sunlight is a good method of bleaching the bags in many cases.

When labels do not respond to the above outlined treatments, soak the bags overnight in kerosene or turpentine, then wash in soap and water. If any color remains, treat them with a commercial bleach. Always remember which treatment works best with certain kinds of bags and file the information for future reference.

If you are unable to remove all trace of the label, finish the job by dyeing one of the darker colors that will cover up the last trace. Dark brown, navy blue, dark gray, green or black are deep enough for this. Black is the only one which will completely cover a black ink label.

If you want to use bag material for lingerie, use the light-weight material and dye in lovely pastel shades such as pale pink, yellow or blue. Heavier bags may be used for coveralls, work smocks, skirts and slacks. In-between weights of material are good for draperies and curtains, pillows and table linens.

Trimmings of all kinds add a dressed-up, finished appearance to any garment, especially if cotton bag material is used. Select it with as much care as you do the colors of your dress.

Ironing Problems

The way your dress looks after laundering depends a great deal upon the ironing technique you employ. Here is the approved order. Iron sleeves first, then the blouse. Then turn to the skirt and iron the hem up and down, never across the width of the hem. Iron collar and trimmings last.

Pleats: Iron the hem and the skirt on the wrong side first. Put the pleats in on the right side with a pressing cloth over the fabric to prevent shine. If there are a lot of pleats, pin or baste them in place, then press. Always iron with the grain of the material.

Shoulders: Use a sleeve board or tailor's cushion to make them neat and smooth. Adjust the cap of the sleeve over the tailor's cushion and press the shoulder as far down as the iron will take it. Avoid poking iron into the seams.

Zippers: Close zipper before ironing, then place a thick towel under the zipper, a pressing cloth over it. This prevents an ugly, shiny ridge.

Fashion Flashes

ALAN LADD
GAIL RUSSELL
in
"SALTY O'ROURKE"
Popeye — News — Cartoon

"ACCOMPLICE"
WANTED BY THE LAW HUNTED BY THE LAWLESS!
with RICHARD ARLEN

Starting Saturday 11 P. M.

THE MUSICAL THAT MAKES YOU FEEL SO YOUNG!
Three Little Girls in Blue
IN TECHNICOLOR
June Haver
George Montgomery
Vivian Blaine
Celeste Holm
Vera Ellen
Frank Latimore

Special for JANUARY only

DANNEN DAIRY FEED
In Attractive Dress Print Bags

• All during January, we're offering Dannen Dairy Feed in attractive dress print bags at no extra cost. You ladies will like these large bags ... takes only two of them to make an entire dress. Smart new patterns.

You'll want to use Dannen Dairy Feed anyway, because it will help you get more milk, and make more profit from your cows. So arrange for your supply now, and get these attractive dress print bags at the same time. Come in and see us today.

DANNEN ELEVATOR
WEST OF THE WATER TOWER

Ask For DANNEN FEEDS

Dress Print Bags? YES, MAM!
SPEAR
FEEDS for LIVESTOCK and POULTRY

Empty Bags Provide Splendid Material for Articles of Home Needlecraft

Housewives are delighted with the attractive sewing material they get every time a bag of SPEAR FEEDS is empty. You can make dresses, curtains, table covers ... scores of useful articles, from this material.

Paper label soaks off easily. A variety of colors and designs. Just another way it pays you to use scientifically prepared SPEAR FEEDS!

HART & SON
Bland, Missouri

NOW! 50 LB. BAGS
Enriched Mother's Best Flour
in lovely DRESS PRINT MATERIAL

For Sale at the Following Grocers:

LINNENBRINK'S GROCERY
J. D. HOMFELDT & SON
C. H. SCHAEPERKOETTER
LICKLIDER'S GROCERY
WEBER BROS.

BLAND MILLING CO., Distributors

DRESS UP YOUR FAMILY AND HOME WITH FEED BAGS!

This young lady has made an attractive dress and smock from Used Kasco Dressprint Bags.

KASCO FEEDS

"BACKED by RESEARCH • PROVED by USE..."

now come to you packed in beautiful, durable cotton dress Print Bags. There are many constantly changing beautiful patterns in all shades and colors from which to choose — all tub and dye fast.

Come in — let us show you these beautiful bags and tell you about the DOUBLE VALUE which KASCO FEEDS OFFER.

There is better feeding results in every bag of Kasco Feeds.

Kasco Feeds are sold by

HENRY E. LANDIS
Abbottstown, Pa.
Phone East Berlin 26-R-6

D. H. SHARRER & SON
New Chester, Pa.
Phone New Oxford 116-R-2

C. M. WOLF
Gettysburg, Pa.
Phone 30

E. L. CARR
Manchester, Md., and Maple Grove, Md.

Announcing A BIG EVENT IN OUR HISTORY

Our Store is Now Headquarters for the Complete Line of

Nutrena
LIVESTOCK AND POULTRY FEEDS

Yes, we have them now, folks ... Nutrena Feeds. We have a complete line for your every feeding need. We are proud to join the Nutrena organization and bring you this outstanding line of feed. With Nutrena, we know our customers will now get the benefits of the latest nutritional discoveries. Won't you come in and let us tell you about the tested Nutrena Feeding Programs. There

is one to fit your operations.

Nutrena Feeds come in economical 50 pound paper bags or beautiful dress print bags. Remember, too, you can buy Crumblized Nutrena Poultry Feeds at no price premium over the mash form. Come in and join the thousands who go after more feeding profits with Nutrena.

Seney Grain Co.
5 Miles North On Hwy. 33
Phone 39F030 Le Mars, Iowa

OMOR *Wonder* ENRICHED

The Winner of Over 16,000 Baking Awards

BEMIS BAND-LABEL BAG

50 Lbs. Net Wt. BLEACHED
The flour in this sack was milled from selected wheat and is guaranteed to satisfy your needs for unexcelled baking performance. Your better baking is insured by a positive money-back guarantee.

VALUABLE COUPONS INSIDE

OMAHA FLOUR MILLS CO.
OMAHA, NEBRASKA

NO BETTER FLOUR AT ANY PRICE

Various plain and solid colour sacks for flour and feed.

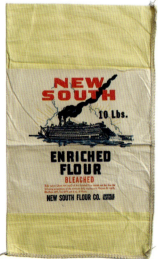

The WANT ADS

ANNOUNCEMENTS

SPECIAL NOTICES 1-A

NOTICE

) whom it may concern, trespassing allowed on y property, south of arroll.

Wm. H. Simons

1A-107-3tp

LOST 2-A

EMEMBER MOTHER AL- vays and save by buying the rift at Charley's Place. 5A-107-3tc

PERSONAL 4-A

BUSINESS OPPORTUNITIES

LOANS 2-B

ANS: YOU CAN OBTAIN A oan for any purpose, quickly at our local office. When buying on the installment plan, ask your dealer for our plan. Tele- phone 253. Office 212 Masonic Temple Building. S&M Finance Co. 2B-101-tfc

BUSINESS SERVICES

INTING & DECORATING 1-C

PAINTING AND DECORATING

AUL E. PATRICK

hone 80 Glidden, Iowa

Number Worth Calling

Free Estimates

1C-108-3tp

Spray Painting

'e are now contracting ray paint jobs. We will ave a limited amount of hite house paint. Call -W for free estimates.

GAMBLES

1C-52-1tc

TRANSFER & HAULING 2-C

OTICE: REGULAR GARBAGE pick-up. Phone 950-W. Ed Haga- man. 2C-89-26tp

REPAIR SERVICE 3-C

EWING MACHINES REPAIR- ed. I have service and part books on all makes of sewing machines. Don't pay traveling racketeers any high prices. Ben Steen, Phone 742. 3C-105-6tp

UARANTEED WORK ON home and auto radios. Record changer repair. The Service Shop back of Galloway Grocery. Phone 1060 for pickups. 3C-109-3tc

MISCELLANEOUS 7-C

RAGLINE WORK, DITCHING and basement work. Norbert Truhe, 1741 North Carroll St. 7C-103-26tc

EW ROOFS, OLD ROOFS RE- paired. Flat roofs, decks, shin- gles, roofs resurfaced, patched, coated, weatherproofed. Prices reasonable, work guaranteed. Phone 995 or 969-W for free estimate. C. D. Andrews Roof- ing Co. 7C-59-tfc

PHOLSTERY, MAKING MAT- tresses, and cabinet work. Nel- son Woodwork. 7C-167-tfc

NOTICE

ew Gravel Washing Plant now operation. We can now furnish u with the following washed aterials:

oncrete mix, concrete sand and ck, also road gravel. Gravel de- vered or trucks loaded at the t.

hone 174-J11 for prices and save your concrete jobs.

SCHUMACHER GRAVEL CO.

Carroll, Iowa

7C-107-3tp

AWN MOWERS GROUND PER- fect, pick-up and delivery. Keys for all locks, steel house and locker keys. Kolb Locksmith, 620 N. Carroll, Phone 1156-J. 7C-108-1tp

ATTENTION

Farmers — Homeowners

See Us Before You Fix Up

Asphalt Roofing

Aluminum Roofing

Insulation

Weatherstripping

Floor Sanding

EMPLOYMENT

MALE HELP WANTED 1-D

VETERANS, HERE'S YOUR chance to travel, if you are be- tween the ages of 18 and 24, and are free to travel — United States, Cuba, Canada and Mex- ico. All transportation furnish- ed, average earnings $200 per month. Apply Thursday after- noon from 3 to 5. See H. E. Blaine, Burke Hotel, Carroll. 1D-107-3tp

FEMALE HELP WANTED 2-D

WAITRESS WANTED, FULL OR part time, good wages. Horse- head Cafe, Westside. 2D-107-3tc

GIRL WANTED FOR HOUSE- work to start June 1. Phone 414. 2D-107-3tc

WANTED: LADY TO WORK half days. Home Laundry & Dry Cleaning Service. 2D-108-2tc

DISHWASHER AND WAIT- ress wanted. Day work, good pay. Alice Cafe. 2D-109-3tc

MEN OR WOMEN WANTED 3-D

MEN OR WOMEN WANTED TO make 7-day traffic check, city of Carroll. Contact W. J. Judge, city engineer. 3D-109-2tc

FARM PRODUCTS

HATCHERIES 3-E

MAY CHICKS

We have about 15,000 chicks yet to sell during May. It won't be long until the chick season will be over, so place your order and be sure to get the kind of chicks you want when you can use them best. We will have some Leg-Hamp and Leghorn cockerels, also pullets on order. Feed, bedding, brooder stoves, equipment, Green Gable Houses. Visitors welcome. Closed Saturdays.

LAKE CITY HATCHERY

Lake City, Iowa

Phone 323-2

3E-109-1tc

FARM MACHINERY 4-E

Surge

DAIRY/FARM/EQUIPMENT

DAIRY FARM EQUIPMENT CO.

V. K. Rath

Sac City, Iowa

4E-93-26tc

FOR SALE: TWO GOOD USED mowers and one good used power mower. Charley's Place. 4E-107-3tc

FOR SALE: SINGLE ROW Mc- Cormick-Deering corn picker, good shape. Gail Owens, Glid- den. 4E-108-2tp

FOR SALE: NO. 229 McCOR- mick-Deering tractor cultivator. W. J. Heires, 1 mile north on 71. 4E-108-4tc

FOR SALE: 999 JOHN DEERE corn planter with 120 rods of wire. Ready to go to work. Al Kerkhoff, Templeton. 4E-108-2tc

FOR SALE USED MERCHANDISE

Perfection Portable Milker. Price $90

John Deere steel elevator 40 feet with steel hoist and truck. Price $290

John Deere 8 foot binder for wind rowing or binding. Price $65

One steel wheel wagon gear. Price $45

Portable Easy Milker. Price $90

WARDS FARM STORE

4E-107-3tc

SEEDS & FEEDS 5-E

FOR SALE: FUNK'S "G" HYBRIDS, early and medium maturity, consistently good, on hand at all times, see Nick M. Wit- try, Arcadia, Iowa. 2 miles northwest of Maple River. 5E-94-30tc

FOR SALE: LINCOLN SOY beans. Germination 96%. $4.50 per bushel. Wm. S. Thieleke, Carroll, Iowa. 5E-106-6tp

FARM PRODUCTS

SEEDS & FEEDS 5-E

GOOD NEWS

Sargent MultiRICH Minral Meal

Now in Dress Print Bags

Your favorite hog supplement, to help save corn and feed out hogs fast—Now in the new, popular dress print bags — the finest prints available at ANY price, high tensile strength; all the flattering new colors and patterns. Yes, more reason than ever for using Minral Meal — for EXTRA value.

FARMERS GRAIN & LUMBER CO.

5E-109-3tc

LIVESTOCK

FOR SALE 1-F

FOR SALE: PUREBRED HAMP- shire fall boars. Roy Nelson, Glidden. 1F-103-12tp

FOR SALE: REGISTERED PO- land China boar, reasonably priced. Marvin Hass, 4 miles west, 2½ south of Manning. Botna, Iowa, Rt. 1. 1F-109-3tc

WANTED 3-F

WANTED: ABOUT 20 HEAD OF cattle in an extra good pasture, by the head. Marvin Hass, Botna, Iowa, Rt. 1. 2½ miles south of Manning. 3F-109-3tc

MERCHANDISE

HOUSEHOLD GOODS 1-G

FOR SALE: USED COOLERA- tor ice box; good used enameled kitchen range. Mattes Hard- ware, Dedham. 1G-107-3tp

ATTENTION HOUSEWIVES

Are you tired of dishwashing? New Kaiser dishwasher now avail- able. No motor to wear out, just connect to hot water pressure and drain. Washes and dries in 5 min. Chassis or cabinet models from $100.00 up. For demonstration call or write.

CLARK MOTORS

Kaiser-Frazer Dealer

Phone 39 or 119W

Coon Rapids, Iowa

1G-109-6tc

FOR SALE: 50-LB. CAPACITY ice box in good condition. Phone 855. 1G-109-1tc

SEE THE BEN HUR DEEP FREEZE 12½ Ft.

Price $385.00

JOHN JUERGENS IMPLEMENT CO.

1G-109-3tc

MISCELLANEOUS 3-G

ALUMINUM ROOFING

GAMBLES

3G-107-tfc

STOVERS CANDY

For Mothers Day Gifts

WITTE'S NEWS STAND

3G-108-3tc

FOR SALE: COOLERATOR, like new. Capacity 75 lbs. M. L. Mills, Phone 685-LW. 3G-109-3tc

MISCELLANEOUS

WANTED TO BUY 1-H

WANTED: CLEAN COTTON rags at Times Herald Office. H1-tf

WANTED: OLD IRON SAVING banks. Will pay $1.00 each. R. J. Dolezal. Phone 600. 1H-108-3tc

FOR SALE 3-H

FOR SALE: CHICKEN HOUSES 10'x20', 8'x12'; garage, 10'x18'. 817 Salinger Ave. 3H-106-3tc 3H-106-3tc

FOR SALE: 6 WINDOWS 31"x 51"; china closet; Admiral ra- dio, good condition. Phone 348. 3H-107-3tc

FOR SALE: 1½-HORSE ELEC- tric motor, like new. J. B. Tap- horn, Roselle. Phone 907-J11. 3H-107-3tp

FOR SALE: NEW ZEALAND white rabbits, bred does, some checkered giants and hutches; also set Concord harness. J. J. Sapp, 325 N. Whitney. 3H-107-3tp

CHICKEN SUPPLIES: OIL burning and electric chick brooders, baby chick and flock size feeder and fountains, also electric pig brooders. Wards Farm Store. 3H-107-3tc

FOR SALE: TWO WINDOWS 26"x28" complete with double frame with weights and screens. Nick Wittry, Arcadia. 3H-108-3tp

REAL ESTATE

SLEEPING ROOM FOR RENT 1-J

WANTED: TWO GIRLS TO share room in duplex. Write Box B, c/o Daily Times Herald. 1J-107-3tp

HOUSES FOR SALE 3-J

FOR SALE: 8-ROOM MODERN house, 2½ lots. Located in Tem- pleton. This is one of the best houses in Templeton. Anton Hannasch, Phone 551, Carroll. 3J-100-tfc

FOR SALE: 5-ROOM HOUSE, and bath, west of St. Lawrence Church. Immediate possession. Anton Hannasch, Phone 551. 3J-104-tfc

ACREAGES & LOTS 4-J

FOR SALE: VACANT LOT, well located on paved street, close in. Phone 596-LJ. 4J-107-3tc

FARMS FOR SALE 7-J

FARM FOR SALE: IMPROVED 80 acres near Glidden. $200 per acre. J. E. Wilson, Agent, Lanesboro, Iowa. 7J-107-3tc

FARMS FOR SALE

An improved farm of 300 acres, located in Northern Iowa. Good set of buildings. 275 acres of tillable farm land. 25 acres of pasture. Price $125.00 per acre. $9,000.00 will handle.

Also have a few extra good bargains. Some for even less money.

JOHN J. GNAM

7J-108-2tc

MISCELLANEOUS 11-J

FOR RENT: DOUBLE OFFICE rooms. Address Box J, c/o Daily Times Herald. 11J-106-3tc

REAL ESTATE FOR SALE

Apartment house, income $1500.00. Priced very low for quick sale.

3-bedroom, modern, oil heat house on north side, 3 blocks from St. Lawrence church.

240 acre farm, improved, 3 miles from Adair, $85.00 per acre.

80 acre farm, improved, close to Templeton.

Hardware store in county seat town. Good volume store. Very clean stock.

ROY J. BURNS

Broker

F. J. HALBERG

Salesman

11J-109-3tc

AUTOMOTIVE

USED CARS & TRUCKS 1-K

FOR SALE: 1929 ESSEX 4-DOOR sedan, 22,000 actual miles, good tires, engine, clean, reasonable. S & S Service Station, Temple- ton, Phone 83. 1K-106-3tc

FOR SALE: 1934 FORD TRUCK, new motor. McCoy Motors. 1K-109-3tc

FOR SALE: 1941 CHEVROLET 2-door, good tires, clean and in good condition. Price $1000. D. C. McClue, Lanesboro, Iowa. 1K-109-3tc

PLUMLEY MOTOR CO.

Few Good Used Car Buys

General Auto Repairs

Parts and Service

AUTOMOTIVE

USED CARS & TRUCKS 1-K

FOR SALE: 1938 FORDOR Ford deluxe. Write Box T. Daily Times Herald. 1K-107-3tp

USED TIRES 3-K

WE TRADE TIRES

GAMBLES

3K-107-3tc

FARMERS: FOR YOUR WAGON or trailer get your 650x15 good used tires, tubes and rims at Booth Home & Auto Supply, 2 doors East of Carroll Theatre. 3K-107-3tc

MISCELLANEOUS 4-K

WINKER'S SERVICE — RADIA- tor, body repairing and paint- ing. Also buy and sell used cars. 4K-104-tfc

Legal Notices

NOTICE OF SPECIAL ELECTION

TO: THE ELECTORS OF THE TOWN OF DEDHAM, IOWA:

You will take notice that at a special meeting of the Council of the town of Dedham, Iowa, a special election was called to be held at the town hall in the town of Dedham, Iowa, on the 15th day of May, 1947, to submit to the voters of said town the following public question:

"Shall the town of Dedham, Iowa, construct, equip and maintain a new town hall and contract indebtedness for such purpose not exceeding $4,-500.00 and issue bonds for such purpose not exceeding $4,500.00, and levy tax annually upon the taxable property in the town of Dedham, Iowa, not exceeding six mills per annum for the payment of such bonds and the interest thereon?"

You will further take notice that at said election, the polls will be open from 8 o'clock a. m. until 8 o'clock p. m., and that at said election the ballot to be used will substantially comply with the requirements of the law relating to special elections to pass upon public questions.

J. T. STANGL,
Mayor of Dedham, Iowa.
WHITE & BRUNER,
of Carroll, Iowa,
Attorneys for Town of
Dedham, Iowa.

(April 24; May 1, 8, 1947).

NOTICE—PROOF OF WILL
In the District Court of Iowa,
In and for Carroll County.

STATE OF IOWA, Carroll County, ss.
TO ALL WHOM IT MAY CONCERN:

NOTICE IS HEREBY GIVEN, That an instrument in writing, purporting to be the last will and testament of Wm. Helmkamp, Deceased, was this day produced, opened and read by the undersigned, and that I have fixed Monday, the 26th day of May, 1947, at 9 o'clock a. m. at the courthouse in Carroll, Iowa, as the day for hearing proof in relation thereto.

WITNESS my official signature, with the seal of said Court hereto affixed, this 30th day of April, 1947.

M. J. THELEN,
(SEAL) Clerk District Court.
(May 1, 8, 15, 1947).

NOTICE—PROOF OF WILL
In the District Court of Iowa,
In and For Carroll County

STATE OF IOWA, Carroll County, ss.
TO ALL WHOM IT MAY CONCERN:

NOTICE IS HEREBY GIVEN, That an instrument in writing, purporting to be the last will and testament of Mary Ritzenthaler, Deceased, was this day produced, opened and read by the undersigned, and that I have fixed Monday, the 26th day of May, 1947, at 9 o'clock a. m. at the courthouse in Carroll, Iowa, as the day for hearing proof in relation thereto.

WITNESS my official signature, with the seal of said Court hereto affixed, this 30th day of April, 1947.

M. J. THELEN,
(SEAL) Clerk District Court.
(May 1, 8, 15, 1947).

NOTICE TO CONTRACTORS CALLING FOR BIDS TO CURB & GUTTER CERTAIN STREETS

Sealed bids will be received by the City Council of the City of Carroll, Iowa, at the office of the City Clerk until eight o'clock P. M. May 19, 1947, for the construction of concrete curb and gutter on both sides of Sixteenth Street, between the east line of Simon Avenue and the West line of North Crawford Street, and North Crawford Street, between the North line of Fifteenth Street and the South line of Sixteenth Street.

The amount of work to be done in the construction of said curb and gutter will consist of approximately 1,445 2/10 lineal feet.

All bids must be accompanied in a separate envelope by a check, certified by a bank in Iowa, payable to the Treasurer of said City, in the amount of ten per cent of the total bid submitted which certified check will be held as security that the bidder will enter into a contract for the construction of the work and will furnish the required bond, and in case the successful bidder shall fail or refuse to enter into contract and furnish required bond, his certified check may be retained by the City as agreed and liquidated damages.

Payment to the contractor shall be made by improvement certificates or bonds as the case may be, on completion and acceptance by the City Council. The payment shall be for furnishing all materials, equipment, tools, labor and incidentals for the completion of the same.

The improvements above described shall be completed on or before August 1, 1947.

The contractor shall give bond in a sum equal to the contract price, on the form to be supplied by the City, obligating the contractor and his bondsmen to the faithful performance of the contract, and to punctually pay all laborers employed on said work, and all persons furnishing materials therefor.

All bids must be made on blanks furnished by the City Clerk, and each bidder will be required to state his price per lineal foot of curb and gutter as per the plans and specifications pertaining to the work.

The City reserves the right to reject any or all bids. The contract will be let to the lowest responsible bidder, unless all bids are rejected, or the council for cause shall find it necessary to throw out one or more bids.

Copies of the plans and specifications for doing said work are on file with the City Clerk, at the City Hall in Carroll, Iowa, and with the City Engineer at the City Hall, at Carroll, Iowa, and may be examined at any time by interested bidders.

All bids will be opened in open council at a meeting to be held at the council chamber on the day and hour above fixed and contract will be awarded at that time or such subsequent time to which the council may adjourn.

If a contract is awarded the successful bidder shall execute a contract with the City of Carroll, Iowa, which contract shall incorporate the conditions and covenants above set out and other necessary covenants and conditions necessary for carrying out and completing the curb and gutter work above described according to the aforesaid plans and specifications for doing said work in a good workmanlike manner.

Published by order of the City Council of the City of Carroll, and dated at Carroll, Iowa, this 28th day of April A. D. 1947.

T. J. KERWIN,
City Clerk.

(May 1, 8, 1947).

SUNDAY, MAY 11 MOTHERS DAY

The one day of every year set aside that all of us might pay our respects to Mother.

LET US SHOW HER THAT WE LOVE HER

Our personal feelings are not enough. We must make some other gesture.

The Finest Way of All to Do This Is to Send Her Flowers

One flower conveys more and says more than 1000 words.

A page from the Thursday, May 8, 1947 edition of the Carrol (Iowa) Times Herald declaring the good news that Sargent Minral Meal is now available in dress print bags.

OLBERDING NURSERY & FLORIST SHOPPE

Plan Legion, Auxiliary Dinners Once a Month

Members of Maurice Dunn Post No. 7, American Legion, met in their hall last night to make further plans for the post's celebration, to be held here June 10 to 14.

A chicken dinner preceded the session. Mrs. Matt Bornback and Mrs. Frank Fister were dining room hostesses.

Wayne Huffman, post commander, announces that there will be a Legion and Auxiliary dinner the last Tuesday in every month. The first of these events will be held Tuesday evening, May 27.

Tony Martin Goes To Jail On Speed Count

WEST LOS ANGELES, CALIF. (AP) — Film and Radio Singer Tony Martin was in jail today, serving out a two-day term imposed when he pleaded guilty to speeding 55 miles an hour on a boulevard last April 26. Municipal Judge Joseph Call sentenced him yesterday.

St. Joseph Gradu Are Spending At Boys Town

Members of the ei graduating class of S school are visiting B Omaha, today.

Taking them on th The Rt. Rev. Msgr. P the Rev. Fr. R. J. Q Booth, Dick Collison, Mrs. O. J. Murphy and Lambert. Sister M. C Sister Joseph Marie a them.

Bernadine Cunningh St. Anthony hospital fo last evening.

Park Ro Sko Pa

TONIGHT

Regular skating fo body and a party s by Jean and Beverly and Donna Nollen.

For health's sake Skate, at this, the tion center of the c

The Style Shop

MOTHERS DAY GIFT HEADQUARTERS

Come early for the finest selection of gifts f mothers of all ages. Gowns, slips, gloves, purse costume jewelry. And Mother will tell you th name of "The Style Shop" on the box adds to th value of the gift, but not to the price.

THE STYLE SHOP

Laurette Burns, Owner

½ Block East of Post Office

Auction Sa

SATURDAY, MAY 1

CATTLE

7 Hereford steers, wt., 750.
4 Shorthorn steers, wt., 650.
3 Shorthorn heifers, wt., 700.
3 stock cows.
11 Hereford steers and heifers, wt., 500 to 650.
8 Hereford steers, wt., 450 to 550.
5 small Hereford calves.
1 Holstein milk cow.
4 mixed dairy heifers.

6 mixed steers, wt. 500.
1 Purebred whitefa years old.

HOGS

1 Poland China boa
2 mixed sows, farro
6 small pigs.

SHEEP

12 ewes and 15 lamb
100 bales alfalfa hay
McCormick-Deering cultivator for H.

FURNITURE

Matched walnut bedroom suite—dresser, chest of dra Beauty rest mattress; Roame deluxe bed springs; combin nut book case and writing desk; platform rocker; porce top kitchen table; hall tree; 9x12 Wilton rug; ironb clothes rack; Bissels carpet sweeper; long mirror (all party, like new); 15 bags of potatoes; 1 dresser; 2 rocki 4 kitchen chairs; 1 complete bed; 1 studio couch.

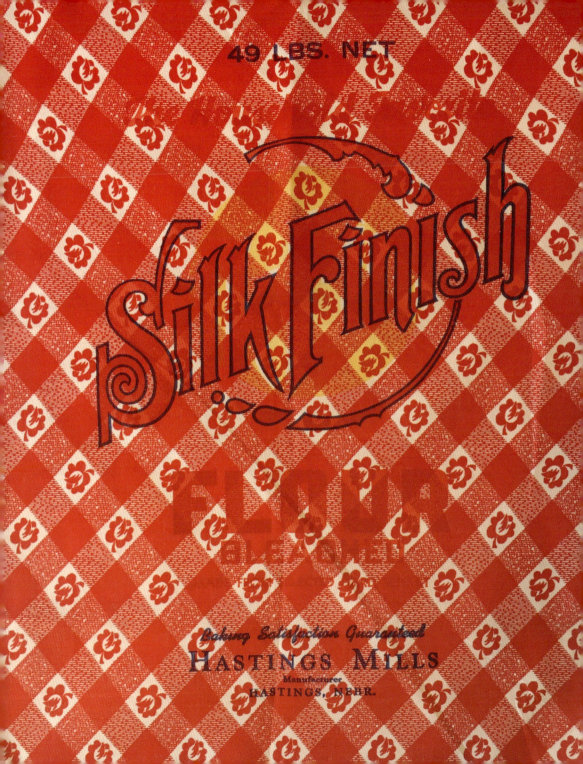

Dec. 21, 1926.

A. T. BALES

PACKAGE

Filed Oct. 29, 1924

1,611,403

GINGHAM GIRL PATENT

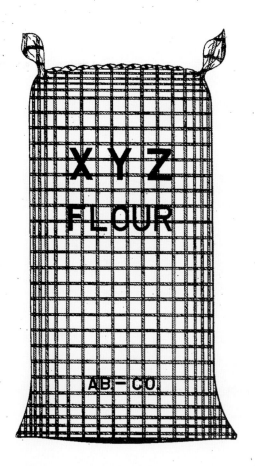

Certain commodities, such as flour, are put in sacks of textile material and such sacks are provided with characters designating the trademark or brand and the manufacturer, jobber or dealer. Such sacks, however, usually serve no useful purpose, except that of holding the product for shipment or until it is used.

One of the objectives of this invention is to provide a package within a sack, the cloth of which can be adapted to be used for dress goods after the product has been removed or consumed.

Inventor.
ASA T. BALES.
John W Brunnga
Attorney

Worth looking into!

Thrifty "home-managers" the country over are fashioning many practical, attractive items for the home and family from Chase-Designed Pretty Print bags. Chase flour, feed and seed bags are available in a large variety of smart, colorful patterns—patterns which assure repeat business and lasting demand for your products. That's why we say Chase-Designed Pretty Prints are "worth looking into." And, we suggest that you do so *today*. Your nearby Chase Salesman will be glad to provide complete information.

CHECK WITH YOUR CHASE SALESMAN ON THESE PRODUCTS
- OPEN MESH BAGS
- PROTEX BAGS—SEWN, ALSO CEMENTED SEAMS
- TOPMILL BURLAP BAGS
- COTTON BAGS FOR ALL NEEDS
- MULTIWALL—AND OTHER PAPER BAGS
- SPECIALTIES

 More than One Hundred Years of Experience in Making Better Bags for Industry and Agriculture.

FOR BETTER BAGS... BETTER BUY CHASE

CHASE BAG CO. GENERAL SALES OFFICES, 309 WEST JACKSON BLVD., CHICAGO 6, ILL.

BOISE • BUFFALO • CHAGRIN FALLS, O. • CLEVELAND • CROSSETT, ARK. • DALLAS • DENVER • DETROIT • GOSHEN, IND.
HARLINGEN, TEXAS • HUTCHINSON, KAN. • KANSAS CITY • MEMPHIS • MILWAUKEE • MINNEAPOLIS • NEW ORLEANS • NEW YORK
OKLAHOMA CITY • ORLANDO, FLA. • PHILADELPHIA • PITTSBURGH • PORTLAND, ORE. • REIDSVILLE, N.C. • ST. LOUIS • SALT LAKE CITY • TOLEDO

PRETTY PATTERNS

Even though people often had no choice but to reuse every bit of what they had, this did not mean that the women who sewed with sacks were satisfied with plain white or off-white clothing and domestic linens. To render them less utilitarian they coloured their feed sack creations with natural or purchased dyes, applied elaborately crocheted edgings and stitched on trims, rickrack and ribbon.

Of the many things manufacturers devised to increase sales of cotton bags (and the items they contained), creating bags of fabric with prints and patterns was undoubtedly the most savvy. Patterned bags offered a touch of fashion with less time and effort. Several companies claimed credit for the idea of offering printed sacks in the late 1930s, but printed sacks existed as early as the mid-1920s. A patent for bags of dress print fabrics was filed by Asa Bales on behalf of the George P. Plant Milling Company of St. Louis in 1924. Bales' vision, however, was limited to one fabric: a red-and-white, small-checked gingham, with a logo printed with vegetable-based inks. A 1925 Gingham Girl flour ad in the trade magazine the *Northwestern Miller* boasted "the remarkable merchandising advantages of this distinctive brand."

Gingham Girl flour was not the only product produced in bags with allover prints. Researcher Ruth Rhoades documented 43 references to printed sacks between 1927 and 1937, including stories from women

An ad for dresses in the June 5, 1924, edition of the Alexandria Daily Times-Tribune *shows the fashion of the times.*

who wore them, samples from Bemis Bag Company and an interview with a Fulton Bag employee who saw them in 1934. These printed sacks were produced in limited numbers and apparently did not catch on in the way that they did starting in the late 1930s.

In 1936, Staley Milling's Tint-Sax bags hit the market. The bags, produced by the Percy Kent Bag Company, were available in nine pastel shades and stitched of a finer-weave fabric. Tint-Sax bags indicated a clear awareness among producers that although women might not be buying the family's feed and flour directly, they were instrumental in making decisions about what constituted value, and how best to spend their family's income.

Cotton bags with dress-fabric designs appeared in the later years of the Depression. Several companies laid claim to being the first to transition sacks from plain fabric to the thousands of prints that were eventually produced. One of the most cited stories is that of Percy Kent Bag Company vice president Richard K. Peek. While dining in a restaurant in Wichita, Kansas, in 1937, Peek took note of the pastel-printed cretonne (an upholstery-weight fabric) slipcovers on the backs of the wooden chairs, and encouraged his employer to create "fashion fabric" bags. By the late 1930s print bags, dubbed "Ken-prints," were appearing in photographs and advertisements, and were available from several companies. A 1947 advertisement in the trade publication the *Northwestern Miller* showed two smartly attired bags walking arm in arm while being admired by passersby. The ad copy assured those considering buying the bags that millers and manufacturers were talking about the strength and adaptability of the bags, as well as "their amazing sales-pulling features." Meanwhile, women were talking about their reusability, "as they bedeck family and home in gay cotton materials."

"Try a Sack Today," suggests a Gingham Girl newspaper ad from 1928.

Illustration from a Percy Kent ad, 1947.

Bag manufacturers placed ads in trade publications, assuring flour mills and seed and feed producers that packaging their product in dress print bags would increase consumer appeal...and sales.

GLORIA HALL COLLECTION

By 1941, approximately 50 million dress goods bags had been manufactured by several companies and sold throughout the United States, with the majority in the South (the location of many bag manufacturers), Midwest and Southwest. A February 22, 1942, *Los Angeles Times* article entitled "Sackcloth Up to Date" noted that three million American farm women and children wore feed sack garments, and that Pacific Mills had created bags in more than a thousand different patterns. Companies continued to make white and off-white bags, but dyed and printed bags remained popular, despite costing five to seven cents more per bag. Around the time printed bags were introduced, manufacturers upgraded the fabric used to make them from coarser weaves to department-store quality cloth.

CONTAINER AND PREMIUM

"All-in-One!"

SELF-CONTAINED Packaging your product in a P/K Bag provides a container and premium all-in-one! There's no bother about enclosures—no delivery or redemption problems. The premium is the package in which your product is packed.

SELF-LIQUIDATING P/K Bags do not represent an additional item of expense. Your product must be packaged. Why not pack it in the popular Ken-Print Bag? It is a package of greater value ... that appeals to consumers because it can be re-used in so many ways. It adds to the satisfaction and enjoyment of your product even after it is used.

PREMIUM-QUALITY Ever since Percy Kent pioneered with the first Ken-Print Bags, women have been clamoring for them. Gay Ken-Prints—Shantung Cambrics—Napkin Bags or Pillow Case Bags. Bargain-wise women naturally prefer merchandise that's wrapped in a premium of attractive cotton material.

P/K Bags are adaptable to a wide variety of products ... perhaps to yours! We'll gladly contribute our ideas. Write us.

From Bag to Napkin by unraveling the seam.

A restyled Feed Bag becomes a dress-up for Sister.

Embroidered design on flour bag makes lovely pillow slip.

"Always Something New"
PERCY KENT BAG CO., INC.
Designers and Manufacturers of Cotton and Burlap Bags Since 1885

Buffalo Kansas City New York

TWO-WAY BARGAIN *in Packaging...*

for You...

The Ken-Print Bag serves as a sturdy, reliable container for your product. Made of premium quality material, high-grade Ken-Print Bags supply a packaging need that cannot be duplicated with less useful containers.

for Your Customer...

The Ken-Print Bag converts easily into "fixin's" for her home, or clothing for her family. Whether it is a fashionable Ken-Print, Napkin, or Pillow Slip Bag, it's a premium of real value that appeals to women. As a premium, the Ken-Print Bag is no added expense.

If you can package your product in bags, why not try colorful Ken-Prints? ... container and premium in one! We'll be glad to make suggestions. Just write.

"Ken-Print Bags Are Worth the Difference in Cost"

PK PERCY KENT BAG CO., INC.

Kansas City   Buffalo   New York

Various ads from 1940s newspapers highlighting dress print bags.

Ads from the Chase Bag Company's monthly trade journal Bagology, *published in February 1953 (facing page) and June 1947 (below).*

In addition to appealing to the "waste not, want not" mentality of women who lived through the Depression, feed sacks provided some women with an actual income. While rural men typically oversaw crops, cows and other large livestock, raising chickens was often the responsibility of women and children. Researcher Ruth Rhoades notes that as well as earning "egg money," Georgia women could make additional cash by selling the sacks that held their chicken's feed. One couple she interviewed raised 80,000 chickens and purchased railroad carloads of 100-pound bags of chicken feed twice a week. The wife, Elizabeth Lott, said she sold thousands of sacks, including 200 to 300 a week to a woman from Southern Georgia. Women with smaller chicken operations could also sell their used feed sacks, which went for around 25 cents each, to chicken-less neighbours or to travelling peddlers, who resold them as they went from farm to farm. Some mills also bought back used feed sacks.

Newspaper advertisements for Gooch's Best Flour, 1938-1940, Nebraska.

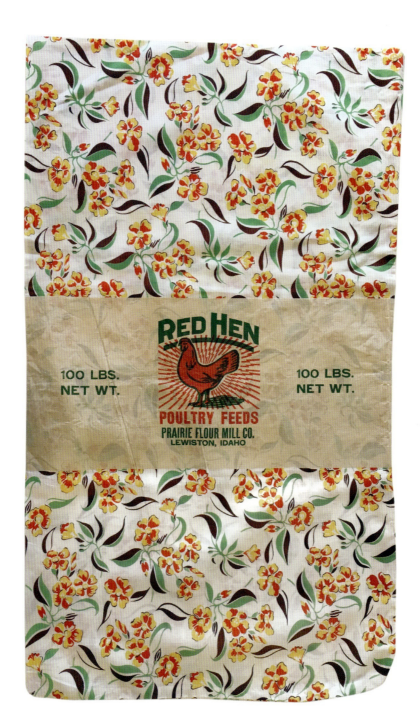

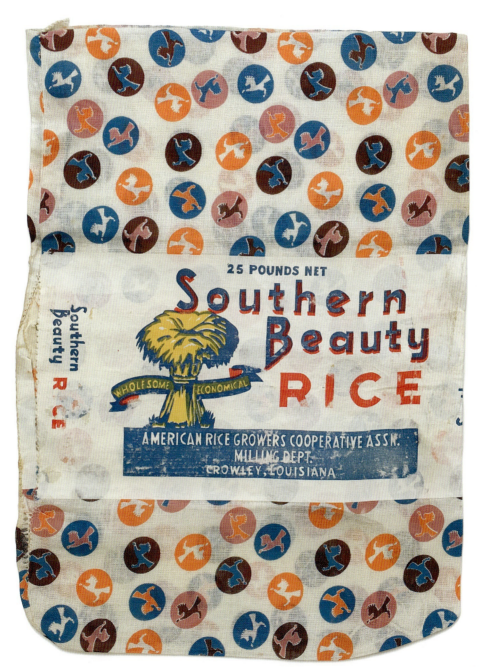

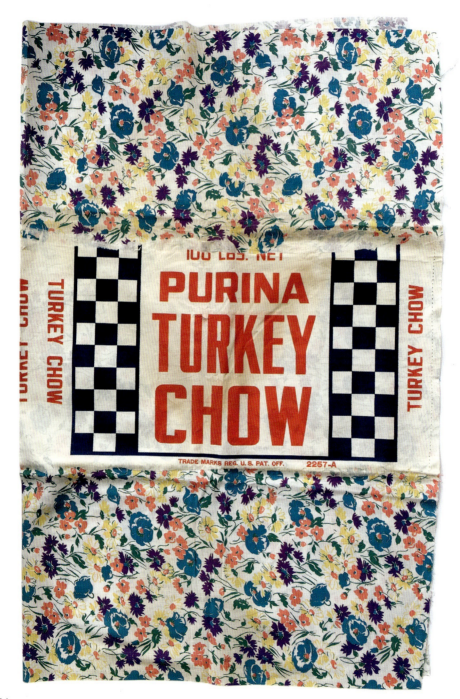

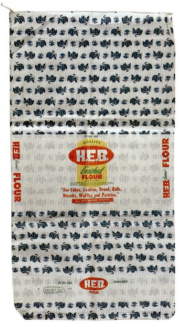

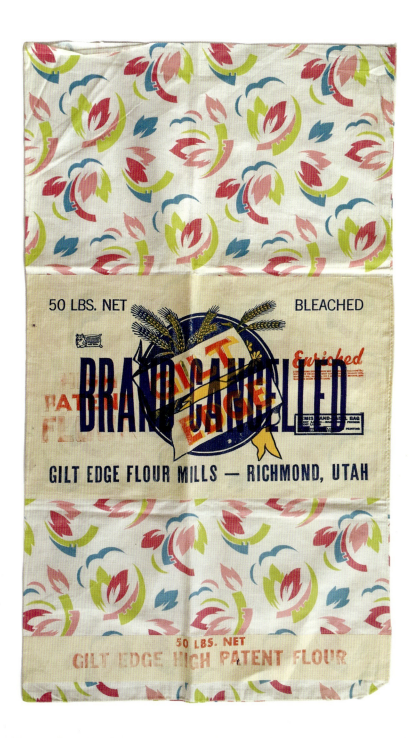

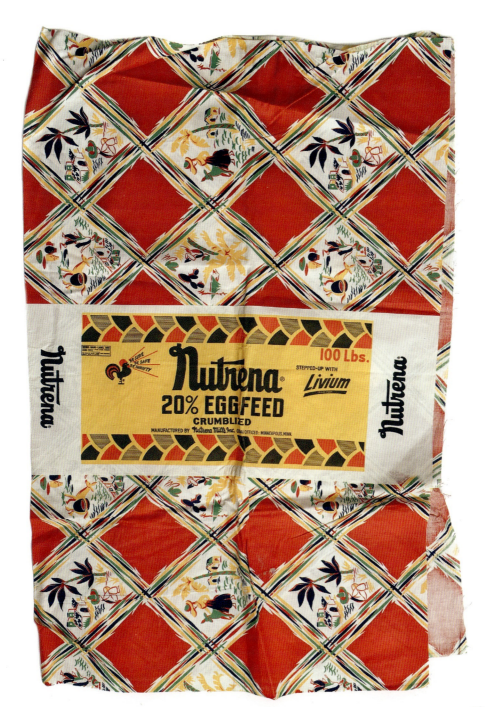

An ad from the Chase Bag Company suggesting that an attractive "package" will sell more product. In this wartime ad, the scenario is about selling war bonds and stamps, but the real message is about specifying Chase Bags.

The ad Boatloads of Bags—Every U.S. Fighting Man is Fed from Cotton Bags Every Day—was part of the newspaper ad series "Cotton Bags at War," produced by the National Cotton Council of America, 1944.

An unsung war hero is "the man on the line where bags are packed with grain that feeds the soldiers who win the war." Published in Bagology, April 1944.

The 60 Million Army of feedstuff bags to supply food-producing farmers. Bagology, June, 1943.

GUESS WHO MADE THE SALE!

If you ever tried to sell a perfectly good product in a dowdy package, you know exactly why Chase customers keep coming back year after year.

Chase not only gives you a fine appearing, well printed package, but it goes way back beyond obvious qualities to fundamental values . . . things you can discover in but two ways. One of these is through laboratory testing. The other, through hard bag-breaking usage. Chase bags take it on the chin and come up smiling.

Why? Because of fine quality, high test fabrics, extra stitches and meticulous craftsmanship. Test Chase bags yourself and see why Chase bags make the sale.

Always specify bags by Chase.

CHASE BAG CO.

GENERAL SALES OFFICES · 309 W. JACKSON BLVD. · CHICAGO, ILLINOIS

WORLD WAR II

Beginning in 1941, shortages created by World War II impinged on the cotton bag industry. Although articles of clothing were not rationed in the United States in the same way that they were in Great Britain and other countries, the majority of United States-produced cotton fabric (along with wool and other textiles) was requisitioned for the good of the troops. A 1946 article in the *Cullman Democrat* from Alabama noted, "On the war front, cotton bags were amazingly versatile—water bags, duffle bags, barracks bags—all made from cotton. Millions of tiny bags packaged silica gel, used by American soldiers overseas. Still other millions of larger bags safely carried foods to servicemen everywhere."

With most cotton going to the war effort, by-the-bolt yardage was hard to come by. Feeds sacks, however, were classified for "industrial" use, and therefore not subject to the same strictures. Burlap was also limited to military uses, and feed manufacturers who used it switched to cotton, increasing the availability of reusable bags on the home front. The range of bag sizes, which had been based on barrels and on manufacturers' assumptions about the needs of consumers, was regulated

A LETTER HOME
APRIL 20, 1945

"The war had developed the art of letter writing. A good letter writer, it appears, is born, not made. The difference between a good letter and a poor letter writer is simple. The good letter writer puts himself or herself in the place of the person to whom the letter is written and tells him the things he wants to hear. They are not necessarily big or important things. Rather they are the little things that happen at home and to the people he knows at home. Little Mary has a new dress made from a feed sack. A piece of the cloth in the letters brings home to the very fox hole of the man who is reading the letter. He sees Mary in the new dress and has visions of how she looks. The young wife wore the dress to the movie last night that she wore the last time he saw her. He remembers it and it brings another picture of home and loved ones. It transfers the little, simple, but wonderfully rich things of home and family across the thousands of miles of ocean. A single page of such letters is worth sheets and sheets of impersonal patter of things, or lonesomeness and worries and headaches. Good letter writing is one of the finest writing arts."

SUFFOLK COUNTY NEWS, SAYVILLE, NEW YORK

Kent's Cloth of the United Nations was a wartime feed sack cloth produced by the Percy Kent Bag Company circa 1944.

GLORIA HALL COLLECTION

by the War Production Board in 1943, streamlining production so that more of the industry's efforts could contribute to victory (see page 359). This had the added benefit of enabling sewists to more easily assess the amount of fabric each bag would provide. (A 100-pound sack produced about a yard and seven inches of fabric, and yardage for women's dresses typically required three to four sacks.)

The necessity of doing things yourself during the Depression meant that the number of women who sewed had increased, and during World War II, sewing was not only encouraged but deemed patriotic. All this helped cotton bag sales to peak during the war years. Women held "bag parties" to swap feed sacks, to ensure that they would have enough bags of the same pattern to stitch a dress or coat. Clothing styles that re-

quired less fabric became fashionable—hemlines rose to above the knee and dress silhouettes became slimmer. "A yard saved was a yard gained for victory" was a slogan touted by the Textile Bag Manufacturers Association. Lighter-coloured fabrics saved on dyes, and cheery design motifs were introduced to help keep spirits high.

Feed sack companies demonstrated their patriotism with images of planes flying in a V-for-victory formation and Morse code motifs. "Kent's Cloth of the United Nations," produced by the Percy Kent Bag Company, featured red, blue and yellow illustrations on a white background, with iconography representing both early members of the United Nations and the war, including Axis leaders Hitler, Tojo and Mussolini in a frying pan labelled "Bad Eggs" and "Keep 'Em Frying."

Cotton bag fabrics lend themselves smartly and economically to limitless decorative schemes in the home. Dainty, easily laundered curtains, slip covers, dresser covers, doilies, luncheon sets, draperies, and many other items can be made readily and inexpensively from bags in which you buy your sugar, flour, meal, feeds. Even wearing apparel can be made from the sturdy cloth from cotton bags—cloth which looks like linen if a bit of starch is added during laundering.

Scores of ways in which the soft, white, long-wearing material from cotton bags can be utilized are explained in the new, illustrated 32-page booklet, "Bag of Tricks for Home Sewing." For your free copy, write National Cotton Council, Box 18, Memphis 1, Tennessee.

There are a multitude of uses for fresh white collars and dickies. Easily and inexpensively made from laundered cotton bags, they lend variety to new dresses and suits and give smart, new touches to last year's suits or frocks.

Make a number of them from cotton bags which come into your home as containers of flour, sugar, or other staples. You will find that this material launders gleaming white and that, with a little starch added, it looks like linen. Cut the collar or dickey the desired size and style and edge it with handmade lace or a ruffle. A medium size cotton bag will furnish ample material.

Numerous conservation tricks which save you money and add cotton freshness to your wardrobe are contained in the new, illustrated booklet, "Bag of Tricks for Home Sewing." Free copy may be obtained by writing National Cotton Council, Box 18, Memphis 1, Tennessee.

Cotton bag fabrics lend themselves smartly and economically to limitless decorative schemes in the home. Dainty, easily laundered curtains, slip covers, dresser covers, doilies, luncheon sets, draperies, and many other items can be made readily and inexpensively from bags in which you buy your sugar, flour, meal, feeds. Even wearing apparel can be made from the sturdy cloth from cotton bags—cloth which looks like linen if a bit of starch is added during laundering.

Scores of ways in which the soft, white, long-wearing material from cotton bags can be utilized are explained in the new, 32-page booklet, "Bag of Tricks for Home Sewing." For your free copy, write National Cotton Council, Box 18, Memphis 1, Tennessee.

It's no trick and no expense to teach the young daughter to sew if you utilize the material from cotton bags. Let her make a wardrobe of doll dresses from the small cotton bags in which you purchase such staples as sugar, meal, and flour. Soon, too, she can learn to make many other attractive and serviceable items and to decorate them with simple embroidery stitches.

The older daughter in school can use laundered material from large

cotton bags in her sewing class.

Illustrating the basic sewing stitches, and offering scores of novel ideas for making useful things from cotton bags, a new illustrated booklet, "Bag of Tricks for Home Sewing," is available to you free. Write for your copy to National Cotton Council, Box 18, Memphis 1, Tennessee.

Practical, serviceable, charming to wear, there's nothing that lends more to the housewife's workday costume than an attractive apron to protect her dress from soil and wear. Every woman needs numerous aprons in a variety of styles and designs.

Aprons can be made inexpensively from cotton bag fabrics which come into your home as containers for flour, sugar, and other staples. Whether it is a large cover-all apron to protect

your dress while cooking, a clothespin apron to serve on wash day, or the dainty apron for guest occasions, the material from laundered cotton bags is readily adaptable. Methods of trimming are countless.

Many sewing tricks, involving fabric-saving, inexpensive ideas are included in the new, ill illustrated booklet, "Bag of Tricks for Home Sewing." To get your copy without cost, write National Cotton Council, Box 18, Memphis, Tennessee.

Sunsuits are a "must" for children. They are inexpensive, too, if made from the sturdy, washable material that comes to you free every week in feed, flour, and sugar bags.

Laundered cotton bags give you an easy-to-handle material, naturally white but easily tinted or dyed

to suit your taste. It is serviceable, comfortable to wear. An embroidered monogram and any one of dozens of ideas may be employed if trimming is desired.

Sunsuits and numerous other money-saving ideas are contained in a new booklet, "Bag of Tricks for Home Sewing," which brings you many profitable sewing tricks with cotton bags. For your free copy, write National Cotton Council, Box 18, Memphis 1, Tennessee.

There's nothing more practical than smocks. They are simple to make, stylish and charming to wear. Use laundered cotton bags to make them, leaving the material in its natural white or tinting it to the color you prefer.

Add a touch of contrasting color at the pockets and neck. Keep several smocks handy. They will save you much time because you can slip one over your dress rather than change clothes to prepare

Bring order out of chaos in your clothes closet by installing a series of containers for shoes, hose, lingerie, and similar items, and by making garment covers to protect your clothing. Cotton flour bags afford an excellent material for these home necessities because the cloth is closely woven, will keep out dust, is washable, and is obtainable without cost. Three bags will make a full-size garment cover. Less material is required for containers for shoes, laundry bags, and smaller handy items.

"Bag of Tricks for Home Sewing," a new illustrated booklet which is offered without cost to American housewives, tells how to make scores of serviceable, attractive things for the house. Get your free copy by writing National Cotton Council, Box 18, Memphis 1, Tennessee.

Cotton bag fabrics lend

Smart Apparel, Household Items Can Be Made From Cotton Bags

By CHERIE NICHOLAS

A YARD SAVED IS A YARD GAINED FOR VICTORY

"Bag of Tricks for Wartime Sewing" was a regularly published newspaper column during the spring of 1944.

"Smart Apparel, Household Items Can Be Made From Cotton Bags" appeared in the Brookshire Times from Texas, May 5, 1944.

IT'S IN the bag, that new smock you need or pretty pinafore you covet, or a sun-suit for little sister. The bag is the same cotton bag that holds your flour, sugar, salt and other such commodities as are packaged in cotton.

One of the most fascinating and thrifty hobbies imaginable is this of creating, not several, but hundreds of attractive and useful household items and articles of apparel from cotton bags. It adds to the fun of converting bags into smartly wearable clothes if you keep a stock of gay rick-rack braid and colorful boil-fast threads and yarns on hand just to give an extra flourish of trimming and embroidery touches to the garments you make. There is no limit to the intriguing accents that can be given to your chic bag fashions, and at such a trifling cost, too. Then too, you can inject real drama into the bag-sewing program by dyeing some of the cotton squares in gay Mexican blues and reds, yellows and purples. Make these up into picturesque peasant dirndl skirts and dresses that are worked out in striking color contrast.

The attractive pinafore centered at the top is made of unbleached bags. You can either hem this apron all around or pipe it with bright colored cotton bias tape-binding such as is available at all notion counters for a few cents outlay. To give it extra fillup, trim it with an applique of flower motifs cut from bright cotton print. You can buy packaged assortments of cutout cotton figures and you'll find them a source of joy as they can be used to trim in so many effective ways.

The dress to the right demonstrates convincingly what smart fashions can be turned out of the unbleached cotton bags. Gay colored accents can be added, such as contrast piping or multi-colored rick-rack also the new green, red, blue and yellow plastic buttons which are being used in rotation. Contrast bodice tops are excellent style and this same model could be made up in this way. Here's where the bags dyed in high colors can be made to yield new glamour. Dye up some of the bags you have on hand, you'll be surprised how effectively they work into the scheme of things.

Best of all is the patriotic spirit you show when you salvage fabrics. The housewife who converts cotton bags into the many useful items they are capable of becoming under the magic of willing hands and minds not only serves herself but conserves essential fabrics for her country.

Under the slogan that "a yard saved is a yard gained for victory," the Textile Bag Manufacturers association has prepared a 32-page booklet "A Bag of Tricks for Home Sewing." This free booklet presents practical ways in which bags can be used to make decorative pieces for home, clothes for the family and very pretty costume accessories.

For the charming dress sketched above to the left in the illustration, the designer uses the bag fabric in its natural color, highlighting it with gaily colorful rick-rack used to trim the low-cut U-neckline, the front opening, the pockets and the sleeves.

The adorable little play dress below to the left with rick-rack trim is pretty enough to set any little girl's heart all a'flutter. The youngster to the right in the little sunsuit and matching bonnet is due for applause at any style show. Note the bolero, the bag and the weskit and the bridge luncheon set, each of which is made of cotton bags.

Released by Western Newspaper Union.

Pleated Skirt

A woman with hips too large for the upper part of her body should not wear a slim straight skirt unless it is cut with a group of deep pleats or soft inconspicuous fullness in center back and front. Such pleats and fullness keep the straight lines on the side but allow the wearer enough freedom in movement, give the illusion of grace and conceal the shape of the thigh. A six-gore cut—and sometimes even a four-gore—may be sufficient to add enough flare all around and fit gracefully.

ags Source Of bulous Dress oth Yardage

30,000 Bales Were Used In 1944 To Manufacture Bags

nother great red-letter season cotton sacks is in the bag. With housewife clamoring for more more commodities packaged ese versatile cloth bags, the ated 750,000,000 yards going their making this year will be sformed from container to wear or household dainty as as Mrs. America can thread needle.

ce 1942 the cotton bag market provided the largest single or cotton fiber. Last year a- the 200,000 bales consumed manufacture of bags would de enough material to make dresses, with all the trim- s, for every woman and girl e United States.

ne was when the cotton bag nothing more than a scullery confined to the kitchen. All the homemaker saw in the sack was the usual batch of its, plus a dish rag or kitchen l. But now this loyal old fami- andby has moved out of the en into the bedroom and liv- room, and lives a glamorous life as a cotton fresh curtain ew frock for milady.

th the salvaging of this bon- aterial, an instrumental fac- easing a tight textile situa- cotton bags played the pri- role in the greatest develop- in home sewing during the years. Often this material, e in quality and looks as the s bought across a pre-war er, was the shortest cut, and times the only cut, to essen- wearing apparel and household

the war front, cotton bags amazingly versatile—water duffle bags, barracks bags, ade from cotton. Millions of bags packaged silica gel, used American soldiers overseas. other millions of larger bags y carried foods to service- everywhere.

w that the textile shortage radually being relieved, the y housewife has no intention andoning her popular wartime ice. The thrift appeal of the with a future will not be out- ed by the changing times, nor magic of sewing with cot- bags lose its fascination for en who enjoy exercising in- ity in their sewing ideas.

illustrate over a hundred and ossibilities in each cotton bag, National Cotton Council has ared a 32-page illustrated book ntitled "Bag Magic for Home ng." Free for the asking, s can be obtained by address- requests to the National Cot- Council, P. O. Box 18, Mem- 1, Tennessee.

ATTRACTIVE AFTERNOONS

8689. JABOT FROCK. Sizes 36, 38, 40, 42, 44, 46, 48, 50 and 52. Size 44 requires 5 bags, 36 x 42.

8730. SEMI-TAILORED. Sizes 34, 36, 38, 40, 42, 44, 46 and 48. Size 38 requires 5 bags, 36 x 42.

8666. ILLUSTRATED ON COVER. Sizes 12, 14, 16, 18, 20 and 40. Size 18 requires 4 bags, 36 x 42.

8753. SHIRRED SHOULDERS. Sizes 36, 38, 40, 42, 44, 46, 48, 50 and 52. Size 44 requires 5 bags, 36 x 44.

Patterns Listed May Be Purchased at 10c Each

for ALL AGES

8680. PRINCESS LINE. Sizes 12, 14, 16, 18 and 20. Size 18 requires 4 bags, 36 x 42.

8743. WASP WAIST. Sizes 12, 14, 16, 18, 20 and 40. Size 18 requires 4 bags, 36 x 44.

8620. BUTTON FRONT. Sizes 34, 36, 38, 40, 42, 44, 46 and 48. Size 38 requires 4 bags, 36 x 44.

Pajamas AND Slips

8600. SMART SLACK SUIT. Sizes 12, 14, 16, 18, 20 and 40. Size 18 slack with blouse requires 4 bags, 40 x 46. Slacks alone, requires 2 bags, 40 x 46. Blouse alone, 4 bags, 22 x 27. Piece under cuffs of slacks.

8451. PAJAMAS AND BOOTS. Sizes 12, 14, 16, 18, 20, 40 and 42. Size 18 requires 2 bags, 4 bags, for trousers. Jacket, 4 bags; 30 x 36. Slippers, 3 bags, 16 x 21.

8593. GOWN AND CAPE. Sizes 14, 16, 18, 20, 40 and 42. Size 18 requires 3 bags, 36 x 44. 3 bags, 30 x 36. Piece both sides, front and back.

8334. SLIM FITTED SLIP. Sizes 14, 16, 18, 40, 46, 48, 50 and 52. Size 18 requires 4 bags, 36 x 44. 2 bags, 26 x 27.

8488. MEN'S PAJAMAS. Sizes 36, 38, 40, 42, 44 and 46 breast. Size 36 requires 2 bags, 42 x 46. 3 bags, 36 x 42.

Pick Cotton and Save
BUY STAPLES IN COTTON BAGS

Flour—Salt
Feed—Sugar—Meal

are examples of products packed in Cotton Bags.
In every household such supplies are bought regularly—and the bags are free. Just ask for the Cotton Bag package.
The larger more desirable bags can be had by purchasing in the larger quantities, the more economical ones. Smaller bags in which foodstuffs are purchased are also made of fine materials and can be used for making the smaller articles described in this book.
BAKERS HAVE COTTON BAGS which they are glad to sell reasonably and the bags are of the larger sizes which cut to excellent advantage.
LAUNDERED COTTON BAGS may be purchased from your Mail Order House or local department store ready for immediate use.
ASK YOUR FRIENDS to save their cotton bags for you, if they do not intend to use them. Frequently these are considered of not much use—but you know they are.

This hand book on the uses of cotton bags has been prepared by the Association as one of its educational projects and is offered to homemakers, sewing instructors, home demonstration agents, club leaders and other interested groups to show what can be done with cotton bags. This is a project approved and cooperated in by the National Cotton Council.

If additional copies of this booklet are desired you may secure them by sending 5c in stamps or coin to The Textile Bag Manufacturers Association, 100 N. LaSalle St., Chicago

Party-Time Prints

Short and long frocks... cotton bag fashions for the holiday mood. You'll find gay party-time prints among colorful containers that package feed, flour and fertilizer.

Left above, the festive look. Scooped neck, sleeveless bodice, full skirt. Sash is drawn through a wide cinch belt. Simplicity 4342 (35¢), junior misses' sizes 11-18. Size 11 takes 3 bags, 39x48.

Center, Simplicity 4440 (50¢), junior misses' sizes 11-18. A dance fashion in full-length splendor. Contrasting trim at plunging neckline matches wide sash. Size 13 requires 4 bags, 39x54. One bag needed for trimming.

Right above, sophistication for afternoon or evening. Note the square picture neckline that dips into a flattering V in the back, and the gored skirt. Simplicity 4347 (35¢), women's sizes 12½-24½. Two bags, 39x54, will make size 14½.

1954 Idea Book for sewing with cotton bags

Illustration from A Bag of Tricks for Home Sewing, *1944.*

GLAMOROUS FASHIONS FROM BAGS

Illustration advertising a Clothing from Cotton Bags Style Show, *September 1949.*

At left, a page from the 1953 Idea Book for Sewing with Cotton Bags *highlighting "gay party-time prints among colourful containers that package feed, flour and fertilizer."*

As the war wound down and as paper bags became more common, cotton bag manufacturers dreamt up new ways to maintain consumer interest in sewing with sacks. In 1947, the Percy Kent Bag Company hired the New York textile designer Charles K. Barton—one of the few designers to be credited—to help bring a little glamour to feed sacks. Barton, who taught textile design at the Moore Institute of Art, Science and Industry in Philadelphia, was touted as "one of America's foremost fabric designers" and "European by birth and education." Though his urbanity was his selling point, his fabric designs were said to have been informed by a visit to the Midwest, where he observed how rural women used feed sacks. In the same year, Bemis offered Bemelin Prints, also supposedly created by chic (though uncredited) New York designers. In the company's ads, the fabrics were proclaimed to be "different, unusual, desirable patterns usually found only in high-priced, exclusive garments." In 1948, Bemis tried another angle, organizing panels of rural women to choose favourites from a range of fabric designs: their choices determined the patterns found on future Bemis bags.

Multiple colourways were created for many dress prints, in which the same design was printed with different combinations of ink colours. Pat Reid, in her notebooks of feed sack samples, includes examples of

Kasco Home Journal, 1942.

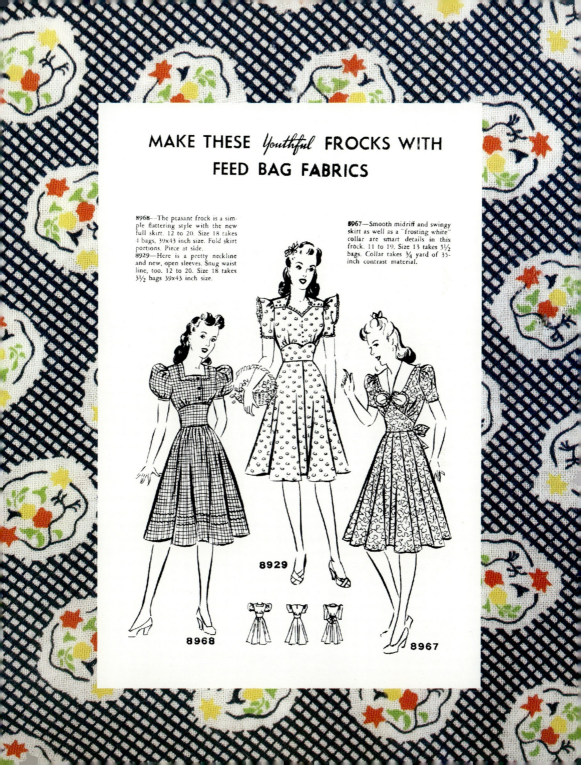

up to seven colourways of a single pattern and presents swatches showing up to a dozen variations of colours. Some of the variety can be attributed to differences in the mix of an ink, the effect of the background fabric on the perception of colour and the discolouration of the printing over time. Feed sack fabrics were produced simply and quickly, and some of the more unusual colour combinations may have resulted when press operators substituted one hue for another when the bags were printed.

Popular culture appeared on feed sacks as another enticement during this era. Licensing agreements meant bags could be adorned with cartoon characters like Li'l Abner and images from movies like *Gone with the Wind*, often printed in multiple colourways. According to author and collector Eddie McGinnis, the colourways were sometimes released one at a time. Around six months after a colourway had been released, presumably after it had sold out, a different colourway would appear, tempting consumers to purchase more. Walt Disney-themed bags, popular for children's wear and featuring Mickey Mouse, Donald Duck and Pluto, were licensed for use in 1950, and Alice in Wonderland appeared after the 1951 Disney movie. Researcher and collector Gloria Hall also has a Disney bag that appeared in 1937 or 1938, in pre-licensing days.

The Textile Bag Manufacturers Association also stepped up efforts to keep cotton bags appealing. The Association joined forces with the

Wayne Chick Starter ad from 1947.

A page from the Kasco Home Journal, *1942.*

Various sewing guides for using cotton bags.

GLORIA HALL COLLECTION

FIRST STEPS
It's easy to get cotton bags ready for sewing...

just rip the bag

and... remove the label!

Cotton bags are sewn with a chain stitch. Cut chain close to bag in corner. Take the top thread in your right hand and the bottom thread in your left, and pull. It rips in a flash!

Band labels are commonly used for brand names. Soak the bag in water and the label comes off immediately! Some brand names are printed on the bags in washable inks that come out easily when soaked in warm, soapy water.

Cotton bags come in a variety of sizes. Use the tables below for determining the amount of cloth in each size.

Millers National Federation (who sold flour in cotton bags) and in 1935 published *Sewing with Cotton Bags,* a pamphlet touting the beauty of bags and suggesting projects for sewing clothing and household items with them. By 1937, 318,000 copies were in the hands of home economics students and teachers, housewives, home demonstration organizations and welfare agencies. (The nonprofit Household Science Institute of Chicago had printed a similar booklet in the 1920s.)

Additional booklets were published in later years, some in collaboration with the National Cotton Council. Thanks to partnerships with pattern companies, these booklets included references to fashionable Simplicity and McCall's patterns appropriate for "mature" and "teenage" wardrobes, as well as children's clothing, and rather than noting required yardage, they listed the number of sacks needed to complete each outfit. The booklets were revised and reprinted a number of times. In 1946 and 1947, more than two million copies of both *Bag Magic for Home Sewing* and *Thrifty Thrills with Cotton Bags* were distributed to sewists.

SLIM as a reed

Presenting... three pattern ideas made to slim, trim, and flatter the figure. *Left*, simplicity in design creates a fashionable dress for any occasion, Simplicity 4016 (35¢), sizes 12-44. Size 40 takes 3 bags, 40-54. *Center*, another becoming frock buttoning diagonally from neck to hem. Simplicity 4087 (35¢), sizes 12-42. Three bags, 40x50, will make size 40. *Right*, the slim tailored look in Simplicity 4017 (35¢), sizes 12½-24½. Size 22½ requires 3 bags, 40x46.

4016   4087   4017

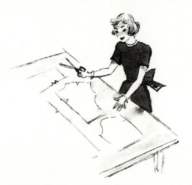

PRINTED FEED SACKS POPULAR AS DRESSES
JULY 18, 1951

The latest fashion news is straight from the feedbag.

American women for years have been turning feed bags into dresses and the manufacturers have helped them by putting attractive printing on the "Cinderella sacks."

Now, with cotton prices rising and shortages forecast, more women than ever before are using feedbags for clothing, according to the National Cotton Council.

A cotton bag style expert at the council recently checked several manufacturers on what's new this summer in printed feed sacks. She reports there are more color combinations and up-to-the-minute designs than ever before.

Floral designs, popular for years, are still around, but the variety is greater. Some are small, others large and sprawling. Colors are pastel or the darker shades. There are plenty of polkadot bags, stripes and checks, Mexican prints, and even modern abstract designs.

Colors combined

Color combinations include navy and rust; grey and yellow; dark green with grey and tangerine; desert tan with light brown, red and black; royal blue with sea green; aqua with various shades of grey, and chartreuse with blue.

For the children, there are snappy Mickey Mouse and Donald Duck prints for nursery decorating or sun clothes. You also can find circus designs, tiny stripes and even animal patterns in feed bags.

From one or more of the bags, the seamstress can make anything from a blouse to to a bedspread, with drapes to match. Many manufacturers supply a pattern with each feed sack. The council has compiled a booklet of patterns, available by writing P. O. Box 76, Memphis.

This spring, there are some new wrinkles to do away with needle-and-thread work. One sack automatically becomes a luncheon-size table cover when the seams are ripped. Another a 25-pound flour sack converts into a pert half apron.

PUBLISHED IN THE MEDFORD MAIL TRIBUNE

In *Thrifty Thrills,* cotton bags were touted as being perfect for projects, and economical. In noting their versatility, the brochure compared them to a fairy tale heroine:

> Once the cotton bag, like Cinderella before the ball, lived a lowly life with the most exciting side of its talent hidden. Though a neat packager for flour, sugar, feed, and meal, it was retired to dishrag duty after these products were poured into cans and bins. Then the American woman, with her imagination and skill—like the fairy Godmother with her magic wand—began to transform the sack cloth into beautiful clothes, drapes, curtains, bedspreads, and small household accessories. Now millions of American women are putting their Cinderella sacks to use in hundreds of new, magical ways!

Beginning in 1944, the National Cotton Council and the Textile Bag Manufacturers Association co-sponsored Cotton Bag Loan Wardrobes, which featured 18 stylish items of clothing created from McCall's patterns and stitched from feed sacks, to women's groups, home demonstration clubs and other organizations for fashion shows. Thirty wardrobes toured the country, "bringing fashion inspiration to many communities," according to a 1948 article in the industrial section of the *Gastonia Gazette* of North Carolina.

Advertising touting the appeal of clothing sewn from bags, some of

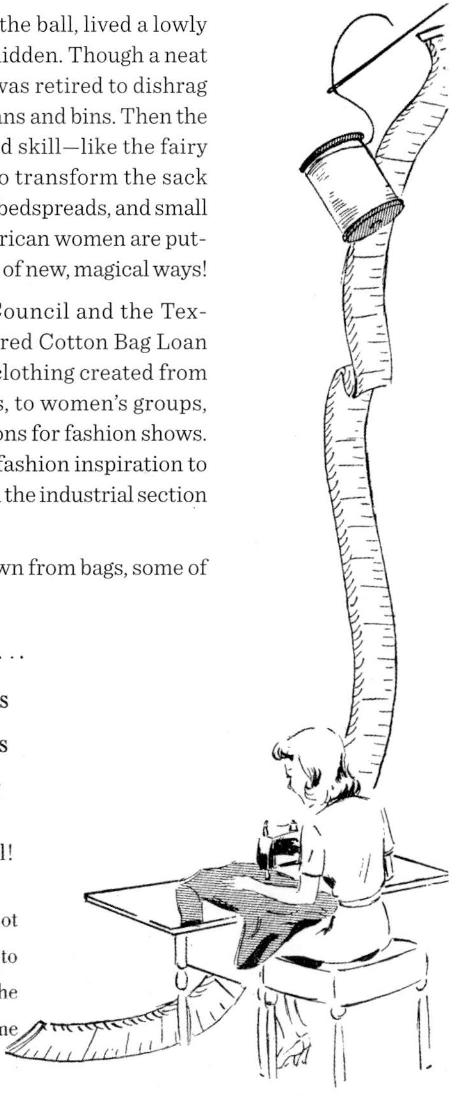

IT FIGURES THIS WAY. IF YOU FEED...

    100 layers .. you get 11 cotton bags

    2 cows ... you get 4 cotton bags

    5 hogs ... you get 1 cotton bag

And that's about 24 yards of sewing material!

If all of the feed, flour, or fertilizer you buy is not available in plain or print cotton bags, be sure to bring this to the attention of your dealer. You're the BOSS! So be sure that the products you buy come in the bags you want.

COTTON BAGS -- Are

That's why women and girls of every age make their special, all-purpose dress from cotton bags

8986 — Teen-age girls adore a combination dress like this which meets all the varied activities in their young lives. 10-20, size 18 jumper requires 2 bags, 36x42 and one bag, 30x34 — (piece front and back skirt). Bolero — 1 bag, 36x42. Blouse, 2 bags, 30x34.

8069 — Here's a dress so easy to make that it could be sewn by the very young miss who wears it. It's in sizes 6-14, and a size 8 girl could make it from only 3 cotton bags, 30x34.

8042 — Whether she has just learned to walk or is grown-up enough for school, little sister can go anywhere in this two-piece frock. Size 4 makes up prettily from 2½ cotton bags, 30x34. In sizes 2-8.

Versatile!

8049

8005

8049—Style-minded juniors rate this their number one frock. It has a wonderfully casual look. Made from 3½ bags, 36x42. Dye a pastel color. Sizes 11-18.

8005—The shirtwaist frock is tops among the tried and true classics. Size 18 can be made from 5 cotton bags, 30x34. Popular in all sizes, from 12-44.

Thrifty Thrills

FEED STORE WARDROBE:

COTTON BAGS ARE USEFUL IN MAKING THINGS

THURSDAY, AUGUST 13, 1953

THE PARIS NEWS, *TEXAS*

Illustrations and headline from the booklet Thrifty Thrills.

COTTON BAGS -- Get All Dressed Up

Everyone agrees that it was really a wonderful day for homemakers when the chickens, the cows and even the pigs got all involved in home sewing!

Thanks to their hearty appetites and to progressive manufacturers who are using fashionable print cotton bags for packing livestock and poultry feed, the housewife has an excellent source of inexpensive fabric for many home sewing purposes.

Today in rural areas, communities, towns and cities, the useful cotton bag has become an important part of American domestic life. Bright, crisp sack curtains lend character to windows in homes from the crossroads settlement to the city apartment. Floral prints and ginghams hob swiftly from the feed store into misses' wardrobes. Plaids and checks adorn the breakfast nook as tablecloths and napkins. Even plain white and unbleached bags undergo a fascinating metamorphosis from containers for products to bedspreads, sheets and slipcovers for home decoration.

According to the National Cotton Council, pretty new bag patterns this spring and summer are certain to inspire the ladies to whip up stylish "sackcloth" additions to their wardrobes.

A preview of the latest cotton bag fashions shows that smartly styled new prints are readily adaptable to fashion trends of the day.

A brown, yellow, green and white check is interpreted in a "Cover Girl" frock—a dapper design with pert collar and cuffs, and pockets that button from waist to hipline.

Homemakers will feel that they are getting the day off to a good start by donning a cheerful red and white print brunch coat, made from three salvaged fertilizer bags. Another easy to hop into in a jiffy is a multi-coloured geometric print. Made from three feed bags, it has coolness features in cap sleeves and cut-out neckline.

An unusually good-looking basic dress is created from two large gray-dyed fertilizer bags. The material has a pretty linen-like appearance and has wonderful durability qualities. A scarf and matching cummerbund made from plaid feed bags may be used to change its personality.

Another outstanding costume is a four-piece play ensemble interpreted in a smart red, navy and white cotton bag print. The shorts and halter are worn for active sports and maximum sunshine, while a skirt and bolero are added for shopping and town wear. Cotton bags star again in suntime fashions with a pastel yellow sundress trimmed in plaid.

Feed bag fashions for children promise to delight youngsters of every age this spring and summer. Among the new designs are a saucy pinafore made from pastel green bags and a checked pinafore cut from red, green, navy and white checked bags. Even sleepytime wear for the kiddies can be made from cotton bags, as illustrated by jaunty blue and white checkered pajamas.

The round-the-clock parade of cotton bag fashions in the spotlight currently includes a southern belle evening dress in soft pink and jet black.

According to the National Cotton Council, many fashions for the men and boys in the family can be made from cotton bags. Pajamas, shorts and work shirts are among the items suggested to win approval of the gents.

which would be considered blatantly sexist today, reached thousands through farm and agricultural trade magazines. A 1947 ad headlined "Nice Package!" featured a smiling blond in a two-piece feed sack sunsuit and a calendar girl pose, and encouraged manufacturers to sell their products in Percy Kent bags. An ad for the National Cotton Council's 1950 *Sewing with Bags* pamphlet included a wasp-waisted, full-lipped woman smiling coyly over her shoulder while pinning a child's dress to the clothesline. The headline read "Local Pin-Up Girl Makes Good," and the text below it added, "she *makes good* dresses, and curtains, and play clothes—and many, many other useful items for the home and family. And, she makes them from plain white and colorful COTTON FEED BAGS."

The women's sections of newspapers ran press releases (often masquerading as news stories) extolling the amazing cotton bag, and an enthusiastic missive in the industrial section of the October 22, 1948, *Gastonia Gazette* equated cotton bags with "the horn of plenty," and not-

JOYCE GROSS COLLECTION

Flour-Sack Dress Booms Anew As Cotton Bags Battle Paper

CHICAGO — Cotton sacks for flour and poultry feed are busting out in a new rash of gay print designs as the textile industry fights to keep paper bags from gaining popularity.

Down on the farm and in many a suburb women are saying they never saw anything in the dry goods shops any snappier than the fancy cloth they're getting free today. The cotton-bag trade is pushing hard to keep them talking that way.

It's a snare to keep modern women buying products because of the fancy wraps, which can be turned into dresses, drapes, slip covers and hundreds of other items. Decades ago, sacks were inked with advertising which could be washed out—or nearly out—leaving a piece of off-white goods for tablecloths, towels, napkins and even underwear. Nowadays sacks come with a minimum of labeling and a maximum of bright-patterned material printed with permanent dyes.

The fight is on because paper bags are cutting a growing swath in the packaging industry. Paper holds an undoubted advantage because as just a container it's cheaper. But, retorts the cotton trade, the gap is closed when cloth container can be salvaged for personal or household use.

A dress design actually costs the buyer about 40 cents; two 50-pound paper bags cost 21 cents. Hence, they boast, the farm wife is getting one and one-third yards of goods for 19 cents.

Burlap bags also have been losing ground to paper ones, which are three times as common for flour and feed as they were in 1939.

To stem the trend, the cotton bag trade has fired its latest shot at the non-farm market. It's trying to interest city wives by advertising the fabric advantages of cloth flour bags in metropolitan newspaper. The industry was delighted when a big New York department store in two days sold housewives 40,000 used flour sacks bought from bakers.

State Prison Free 145

BOSTON — The state Correction Department has announced that 145 prisoners were released today from state prisons. Under an act of the 1948 Legislature, prisoners are let off earlier for working especially hard while behind the bars.

Those released include 29 from Charlestown State Prison, 34 from Norfolk Prison, 31 from the state farm at Bridgewater, 40 from the Concord Reformatory, and 11 from

Modern Barter System Finds Feed Bags Valuable as Cash

If magazine and newspaper swap columns are any indication, owning empty cotton feed bags is just like having money in the bank. Some homemakers may have antique vases and even crystal-footed cake stands to barter but nobody can strike up a better bargain than the trader with extra feed sacks.

Making a check on thousands of items women have offered to swap for salvaged cotton sacks to be used for home sewing, the National Cotton Council reports that housewives holding the bag in these spirited trades possess a magical medium.

For example, a flower enthusiast could practically cover all her gardening areas with rare plants and beautiful bulbs offered in exchange for print sacks. Purple African violets, purple gloxinia rooted leaves, red gloxinia leaves, tall red canna roots, gladiola, iris, and dahlia bulbs are among the many flowers offered.

In the baby department, pink carriage robes, a child's folding car set, and baby buggy have been put up for cotton bags. A housewife from the mid-west wanted to trade a car heater for white feed sacks, plus six small and three miniature flower pots as a swap for two matching print bags.

Make Smart Fashions for Pennies

Summer clothes to make inexpensively from cotton feed bags include "sack separates" with blouse in neat check and skirt in solid color. Pedal pushers (center) were made from two feed bags dyed charcoal gray while weskit came out of one plaid sack in a red, gray, navy and white. Halter dress (right) evolved from three feed bags in yellow, brown and white stripe while single pastel bag was used for bolero. These clothes are made from pattern and at a cost that's truly kind to the budget. A single 100 pound cotton bag contains about one and a third yards of fabric that can be used for many home sewing purposes.

ed the recent changes to what were once "industrial cotton containers, completely lacking in color!" The article added that manufacturers had recognized that women were reusing plain sacks and "answered every woman's quest for glamour by printing their bags in colors that rival the rainbow! All this happened in the last ten years. ... Until today consumer demand, plus the durability and service features of cotton bags, have made them the largest single industrial user of raw cotton."

For a time, feed sacks were so popular that it was possible to purchase them—used and unused, bleached and unbleached—from urban department stores, such as Macy's in Manhattan, and through the Sears and Montgomery Ward catalogues. The National Cotton Council sold bags for women who were having difficulty finding matching bags locally for projects requiring additional yardage. Bakeries also sold their empty flour bags to customers—an October 22, 1948, article in the *Gastonia Gazette* noted, "what began as a rural thrift measure has become a city fashion note. Large bakeries, who receive their flour in the 100-pound bags so desired for their yard and seven inches of material, furnish the urban housewife with her bags. These are sold either over bakery counters, or by bakery door-to-door salesmen." By 1947, an estimated 750 million yards of cotton fabric went toward bag production, accounting for eight percent of all cotton textiles produced in the United States.

Despite these efforts, cotton struggled in the decade that followed to compete with other materials. An article in the November 1, 1953, edition of Phoenix's *Arizona Republic* entitled "Sacking is Promoted as Decorative Fabric" noted that cheap paper and burlap bags were testing cotton's dominance. The yardage consumed by cotton bags in 1953 was estimated at 435 million yards, down from 480 million yards just two years before. A three-ply paper bag with a 50-pound capacity cost around 7 cents, and burlap was 9 cents a yard—class B cotton sheeting, meanwhile, cost 17 cents a yard (it took about 1.5 yards to make a 100-pound bag). In an August 21, 1960, article in Ohio's *Times Recorder* entitled "Exit Flour Sack, Once So Useful," the president of Seaboard Allied Milling was quoted as saying, "Flour for large consumers is destined to be handled more and more in bulk in special trucks and in paper containers for the housewife."

TWO-TIMER . . . Two gray-and-white striped cotton bags make blouse, two plain bags are dyed gray for the skirt. Made from McCall Pattern 8657.

Above, fashion ideas published in "Fashion Steps Out of the Feed Bag," the Evening Sun, May 20, 1952.

Clippings from the Berkshire Eagle, *August 31, 1950;* the Denton Journal, *August 31, 1951;* and the Kingston Daily Freeman, *July 29, 1953.*

Above, This Month in Rural Alabama, *June 1945, Montevallo, Alabama, and at right, various clippings highlighting dresses made from sacks in 1956, 1959 and 1961.*

COTTON BAG CASUALS

This leisure time ensemble was made from cotton containers which package feed, flour, and other staple farm products. Two sacks in a splashy modern print went into the overblouse, and two bleached cambric bags were dyed black for the fancy pants. The outfit was made from Simplicity patterns 1088 and 1059. Cotton bag wardrobes of summer fashions styled by McCall and Simplicity patterns are available for special shows from the National Cotton Council, P.O. Box 9905, Memphis 12, Tenn.

EVENING ELEGANCE — It's hard to believe that this glamorous evening gown is made from cotton flour sacks. The pattern, originally designed for the 1959 Miss America, is McCall's 4870. For information of cotton bag fashions for style show use, write the National Cotton Council, Box 9905, Memphis 12, Tenn.

STRIPES FOR SPRING — Here's a dress appropriate for spring-through-fall wear. The fabric, something new in cotton bag material, has the texture and appearance of linen. The style shown is McCall's Pattern 5369. Cotton bags, used for packing staple products, offer a variety of sewing fabrics.

THE Cotton Bag

STEPS OUT!

FASHION SHOW SUGGESTIONS • FEATURING
COMPLETE COTTON BAG WARDROBES

SPONSORED BY
NATIONAL COTTON COUNCIL OF AMERICA
BOX 18 • MEMPHIS, TENNESSEE

BEAUTY WEARS A SACK

While feed sack apparel could be fashionable, some still felt stigmatized for wearing clothing that once held fertilizer or seed corn. Indeed, advertisements for feed sack clothing attempted to counter rural stereotypes by featuring glamorous women wearing up-to-the-minute styles, and bag manufacturers hired "New York designers" to create stylish fabric prints.

Still, wearing an actual sack was a ploy publicists seemed to love. A nubile young woman wearing a rough sack lacked subtlety, but was nevertheless a favourite way to juxtapose blossoming beauty with country coarseness. In 1951, photos of Marilyn Monroe surfaced in which she was attired in a burlap potato bag. The reason she donned the form-fitting sack seems to be lost in the mists of time, but one theory is that after a columnist called her "vulgar," noting that her clothes were too tight, studio executives decided to prove she could look sexy wearing anything—even a sack. On June 23, 1958, the *Greenfield Recorder-Gazette* captioned a photo "Some Chick." In it a smiling Diane Gay Austin, aged 18, is wrapped in a feed sack "dress" with a deep décolletage, a baby chick perched on her palm. Diane, it seems, had won the Miss Poultry Princess contest in Atlanta and would be representing Georgia in the Miss Universe contest. Despite her beauty and glamour, she was apparently a country girl at heart.

In a 1956 episode of *I Love Lucy*, Lucy goes on a hunger strike during a visit to Paris in an effort to get her husband Ricky to buy her an elegant gown. When Ricky discovers that their friend Ethel is sharing food with Lucy and that the hunger strike is a ruse, he and Ethel's husband Fred find burlap potato sacks and a horse feedbag and have gowns created from them, telling Lucy and Ethel they are high fashion. The reactions of passersby give the women a clue that their gowns are anything but trendy—and they suffer the stigma of wearing feed sack fashion on the streets of Paris. (In the end, the joke is on Ricky and Fred, when the creator of the gowns is so impressed by Lucy and Ethel that he replicates the gowns in actual dress fabrics for his own models.)

The expression "It fits like a sack" doesn't hold true anymore. Even if it is a sack. Medford Mail Tribune, November 27, 1957.

DRESS FROM PAREE?
NO—JUST A SACK!
MONDAY, AUGUST 9, 1948

You can't make a silk purse out of a sow's ear, but you can make a fancy evening gown out of a feed sack!

Farmers' wives have known this for years but we city folk have just begun to catch on—and feed and grain dealers in Rochester report a real "run" on the feed-bag market.

For a long while thrifty, ingenious farmers' wives have been using the sacks for everything from dish towels to children's sunsuits.

And now one of the most smartly dressed, sophisticated women I know takes great delight in saying when friends admire her svelte red and white print sun dress:

"Oh, it's just an old sack"—which is true, of course, but the "old sack" is far removed from the days when it enclosed 100 pounds of chicken feed!

A number of years ago, I learned, all feed and grain sacks were plain white. Feed dealers noticed that farm women were using the bags for all sorts of household purposes, including clothing.

So dealers decided that if the plain, white bags were good, sacks made of attractive, printed material would be better.

DEMOCRAT AND CHRONICLE
ROCHESTER, NEW YORK

Feed-Bag Fashions

By Lenore Brundige

Believe it or not, the gals are making clothes from cotton feed bags. Yessir, the same kind that you see in a country feed store or a city bakery, the kind that holds chicken feed or flour.

Getting the bags ready for home sewing is a simple operation. Almost all sacks come with new band labels which soak off when dipped in water. The bags are sewn with a chain sitch that can be ripped open easily.

Each 100-pound bag contains about one and a half yards of fabric. So it takes three of them to make an average-size dress. As shown in these photos, there is a variety of patterns suggested by the National Cotton Council. Damask tone-on-tones, batik prints, floral stripes, modernistic prints and other striking designs fit in with up-to-the-minute fashions.

And the new summer styles from feed and flour bags can be made for all occasions around the clock—bathing suits, afternoon frocks, evening gowns, pajamas.

So—take a tip from the feed store dealer or neighborhood baker—if you want a new dress that's a genuine original and a bargain to boot, it's in the bag.

UP TO THE MINUTE in a fashion way is this daytime print. It has a Peter Pan collar, very short sleeves and a flared skirt—all in vivid colors.

IT TAKES FOUR 100-pound cotton bags for this pretty and practical morning frock. It has a flattering square portrait neckline, nipped-in waist.

MIGHTY FANCY, but it's a sack suit. It was cut from two 100-pound flour bags salvaged from the baker. Half the second sack wasn't needed.

DANCE FLOOR DREAM is this chic chicken bag evening gown. Eight feed bags were needed to make this swirling, off-the-shoulder formal.

READY FOR BED, this beauty will sleep in an old-fashioned night shirt made from three feed bags, thus completing a day in cotton bag clothes.

FEED SACK COMPETITIONS

1954
"SAVE WITH COTTON BAGS"
Sewing Contest

Feed sack manufacturers and the cotton industry worked hard to reverse the dwindling popularity of cotton bags. Beginning in 1953, the National Cotton Council partnered with sewing machine manufacturers and the Textile Bag Manufacturers Association to hold feed sack sewing competitions in every American state at county and state fairs, with the winner crowned in Chicago. Contestants were lured by tantalizing prizes, which were listed in newspapers across the country, including the *Medford Mail Tribune* in Oregon, and included a gas range, a television set, radios, a 42-piece set of silverware, electric deep fryers and all-expenses-paid trips. In 1955, an estimated 25,000 seamstresses participated.

The *Paris News* in Texas reported on September 17, 1957, that the national winner that year was Mrs. Alma May, "an attractive, 25-year-old queen" from Steamboat Springs, Colorado. She entered an item in each of the seven categories: kitchen curtains, mother-and-daughter dresses, luncheon sets, quilts, pajamas or lounging outfits, shirts and blouses. Winners of the 1958 competition were feted during a weeklong stay in New York City, while the 1959 winners met Jack Bailey, host of the popular television show *Queen for a Day*, and toured Disneyland as part of their triumphant swing through the Golden State. They were also interviewed on a Los Angeles regional farm radio station, lunched at 20th Century Fox's commissary (presumably with movie stars) and fed the world's largest captive whale at Marineland of the Pacific.

These lucky winners were Mrs. J. Melvin Peterson of Selah, Washington, and Dorothy Overall of Caldwell, Kansas, who had entered numerous sewing competitions during the 1950s and 1960s. Dorothy's pieced and appliquéd bassinet quilt, made of solid blue and solid white feed sack fabrics and depicting the silhouettes of two birds and a flowering lily of the valley, was the first runner-up, and it today is in the collection of the Smithsonian's National Museum of American History. According to their website, Dorothy sewed the quilt on a Pfaff sewing machine she had won in another contest.

Younger sewists were also groomed for cotton bag loyalty through 4-H sewing competitions at county fairs. In the August 14, 1945, *Boone County Recorder* from Kentucky, near the historic headline "Japanese Accept Surrender Terms," is a story about the "Four-H and Utopia Fair" in which a "feed sack dress class open style revue" would be featured at the grandstand in the afternoon. In the same competition in 1950, the prizes in the feed sack dress category were $2 for first place, $1.50 for second and $1 for third.

Cotton-Bag Items Merit Prizes

Modeled, above, are winning entries in the National Cotton Bag Sewing Contest held recently in Memphis, Tenn. Homemaker, center, wears multicolored polka dot dress made by first prize winner Mrs. Fred Sowers, of Winston-Salem, N.C. Model's daughter, left, wears white, starched dress stitched by Mrs. Carl Arntson, Columbia Heights, Minn., third-prize winner. At right shown tailored, tucked blue dress which helped Mrs. Harold Vesta, Niangua, Mo., to win second prize. Other items on table are among the many prize-winning creations, chosen from entries submitted by homemakers.

The Bag A Cotton Bag Sewing Contest, with a $350 sewing machine as first prize, will be held at the Arizona State Fair this year. Just ow what can be done with printed feed sacks, Pat Hall, 17, of Tucson, right, na's Junior Maid of Cotton, models a dress made from four cotton bags. In the tself is Diana Laird, 17, Phoenix, 1953 Salad Bowl Queen. Also present is Cinta, an Aberdeen Angus calf, who had a real party out of the feed sack in the round.

Bunch O' Nonsense

ving Contest nounced by ton Council

y response to the 1955 bag sewing contest indi- that the third annual will set new records in ipation, the National Cot- ouncil announces.

nationwide competition, o all women interested in bag sewing, is sponsored e National Cotton council he Textile Bag Manufac- association with the co- ion of Pfaff Sewing Ma-. Contests will be staged at te and regional fairs from to November, with finals led for November 15 in go.

Prizes

winner at each fair will e a portable Pfaff sewing ne, cash awards and a e at the national cotton sewing queen title and prizes. The queen and unners-up will win, for elves and escorts, all- se paid trips to Chicago, a s stay at the New Morrison and gifts worth more than

h an increased number of ies from women wanting an early start in the 1955 t, the Cotton council ex- 25,000 entrants this year. tries must be made from bags, and are judged for ality, workmanship, ap-

Cotton feed bags in interesting floral print have been chosen for an afternoon fashion by Nell Perkovich, young housewife of Memphis, Tenn. The dress will be part of her entry in the third annual National Cotton Bag Sewing Contest at the Mid-South Fair. Mrs. Perkovich wears a tailored style in red, white and blue plaid, also made from cotton bags. Nationwide bag sewing contests will be staged at 52 state and regional fairs from June to November. Finals will be November 15 in Chicago.

Family Arrives **Courtesy Said "Must"**

COTTON BAG WINNER — Mrs. Orville Snider, of 1911 Coggin St. Petersburg, walked off with the top honors in the cotton bag contest at the Southside Virginia Fair. Here she poses with her prize, a portable sewing machine and a portion of her entries in the Home Arts Department. The sewing machine was presented to Mrs. Snider by Mrs. Coleman W. Woodruff, director of the Home Arts Department.

Everett, Wash. Girl Married To Carroll Franklin Glisson to M

$250 IN PRIZES

Will Be Offered This Year in the

Clothing From Cotton Bags Contest

Held in Connection with the

National Cotton Pickng

Win a Sewing Machine, Cash Prizes and Other Awards
Enter the 1954 "SAVE WITH COTTON BAGS" CONTEST at your State or Regional Fair

HOW YOU CAN WIN...

Mrs. Esther McGugin, 1953 International Cotton Bag Sewing Queen, uses her new Pfaff machine to make a cotton bag sun dress for her teen-age daughter. She is the wife of a Fresno, California, cattle rancher.

Details of the 1954 "Save With Cotton Bags" Sewing Contest will appear in the premium book published by your state or regional fair. (Note: See page 4 of this insert for general rules and entry classifications.)

SAVE your COTTON BAGS and start planning for the contest NOW. Each article must be made entirely from COTTON BAGS, with the exception of trimmings such as rickrack, braid, tapes, ribbons, buttons, etc. You will find that feed, flour, fertilizer, sugar, and other farm and home products come in a variety of attractive cotton packaging.

Use your imagination—ingenuity will count as well as fine workmanship and the finished appearance of your sewing. Select your material carefully for each article. Judges will consider such factors as originality of ideas, adaptability of fabric, quality of workmanship, suitability of trimmings. This Idea Book for Sewing with Cotton Bags will prove a valuable aid in planning your entries for the 1954 contest.

WHAT YOU CAN WIN...

You can win a de luxe Pfaff sewing machine, blue ribbon honors, cash prizes and gift awards at your state or regional fair.

You can win, in the national finals, a free trip and a week's visit to Chicago and such valuable gifts as Pfaff de luxe sewing cabinet, Sentinel 21-inch television set or clock radio and Laundry Queen automatic washer. Bell & Howell will give each of three top finalists a movie camera, projector and film record of the 1954 contest.

Cash prizes totaling $66.00 or more and gift certificates worth $680.00 will be offered at each fair participating in the contest. Gift certificates apply on purchase of a sewing machine or cabinet from your local Pfaff dealer. The sweepstakes winner (person receiving the greatest amount of cash prizes) at each fair will be awarded a Pfaff Zig-Zag sewing machine with the exclusive "dial-a-stitch."

The winning sweepstakes entries from all fairs will be judged in competition to select the 1954 International Cotton Bag Sewing Queen and two runnersup. These finalists will be named at ceremonies featured during the International Dairy Show, Chicago, October 9-16. They will appear on national radio and television programs and, with their escorts, will enjoy a full week of royal entertainment in Chicago.

Sunday, May 9, 1954 — The Arizona Republic,

Tucson Grandma To Vie For Crown

A grandmother from Tucson is a competitor for a cotton queen's crown.

Mrs. H. A. Arnold, winner last November of the Arizona State Fair's contest for articles sewn out of printed cotton feed and seed bags, has a chance to become the 1954 National Cotton Bag Sewing Queen at a contest to be decided in Chicago.

Mrs. Arnold's competitors will be needle magicians who made use of colorful printed cotton feed, flour, and fertilizer sacks to win championships in 42 other state fairs.

MRS. ARNOLD was sweepstakes winner over 186 other entrants in the first "Sew With Cotton Bags" contest ever to be held at the Arizona State Fair. Her championship entry was a matching mother and daughter dress set.

The contest, which will be held again this year at the fair, is sponsored by the Textile Bag Manufacturers Association, Evanston, Ill., whose members make sacks in designs suitable for dress goods, curtains, luncheon sets, or almost any use to which printed cotton material can be put.

Co-sponsors are the National Cotton Council and the Pfaff Sewing Machine Co.

THE SWEEPSTAKES prize is a Pfaff sewing machine valued at almost $300. Mrs. Arnold won last year on a sewing machine she had been using since 1916.

Mrs. Arnold's 1953 dress set is in competition for the 1954 National sweepstakes award because the state fair, held in November, follows the national contest which is held in Chicago in October.

Mrs. G. C. Quick, superintendent of entries at the state fair, pointed out that entrants for the 1954 fair do not have to be purchasers of feed or flour in bag quantities to be eligible. They can be bought at stores which sell feed and seed in small units.

The stores save the sacks and sell them. Information regarding entries can be obtained from Mrs. Quick at the Arizona State Fair.

Winners Little Betty Wesner, 4, Tucson, and her mother, Mrs. E. P. Wesner, pose in the mother and daughter dress set which Betty's grandmother, Mrs. H. A. Arnold, Tucson, made from printed cotton flour sacks.

Dastco — **18th YEAR IN THE GROUND**

Our 18th year of ACID DELINTING cottonseed has helped prepare seed to produce stronger plants, better yield of finer cotton.

DASTCO SINKERS

OTHER WEAVES AND FIBRES

Both densely and more loosely woven fabrics were used for commodity bags and were selected for use based on the bag's contents. Osnaburg, a coarse, low thread-count, plain weave fabric, was used early on for bags with chunkier contents, such as animal feed and agricultural seed. More tightly woven cotton fabrics with higher thread counts, including cambric and sheeting, were stitched into sugar and flour bags. As patterned feed sacks appeared, the quality of the fabrics improved, becoming equal to that of dressmaking fabric sold in stores. Indeed, by the 1940s, some textile mills sold the same fabric to manufacturers of sacks and to those who sold it to customers by the yard, on bolts, in stores.

A 1950s-era ad for Merit Egg Mash or Kurnls announced, "Now in Walt Disney's 'Alice in Wonderland' percale bags." Alice herself runs across the top of the ad on the left, while on the right a happy mother wearing a bonnet and apron (likely made of feed sacks) hands the eggs she is gathering to her cheerful children, who are piling them in an overflowing wire bucket. "Mom gets MORE EGGS … we get swell clothes with MERIT!" says the feed-sack shirted son. Further down the page the finer points of percale are pointed out: "80-SQUARE PERCALE—finest print material: Every bag of MERIT Egg Mash, whether it has Walt Disney characters or some other exciting pattern, is made of *fine quality, smooth-finish, 80-square percale*—the strongest, most wearable print material on the market."

Cotton was the predominant bag fibre, but nylon, flax and kenaf—a coarse fibre typically used for rope and animal bedding—were occasionally used. In 1951 Bemis offered bags they called "Bemaron," rayon bags perfect for creating silky smooth blouses, scarves and slips. Sometime in the 1950s, manufacturers began replacing the cotton string used to sew bags with rayon. Although most feed sacks were sewn from plain weave fabrics (with the weft threads alternately going over and under the warp threads), the Fairmont State College Textile Bag Collection from West Virginia has sacks with a variety of weave structures, including a plain weave with multiple warp yarns held together, creating a stripe effect, and a plain weave bag with a gold metallic yarn woven in at intervals. In *West Virginia Quilts and Quiltmakers: Echoes from the Hills*, author Fawn Valentine notes that this was a favourite of high school girls, who appreciated this bit of bling when sewing evening gowns. She adds that "later bags were available in fashionable fabrics like dimity, pique, broadcloth, and seersucker."

The fabric sample on this page is from a feed sack with an unusual weave structure, from the collection of Gloria Hall.

For reasons unknown, this Frankenstein-like sack was made from two different fabrics and brand labels.

PAUL PUGSLEY COLLECTION

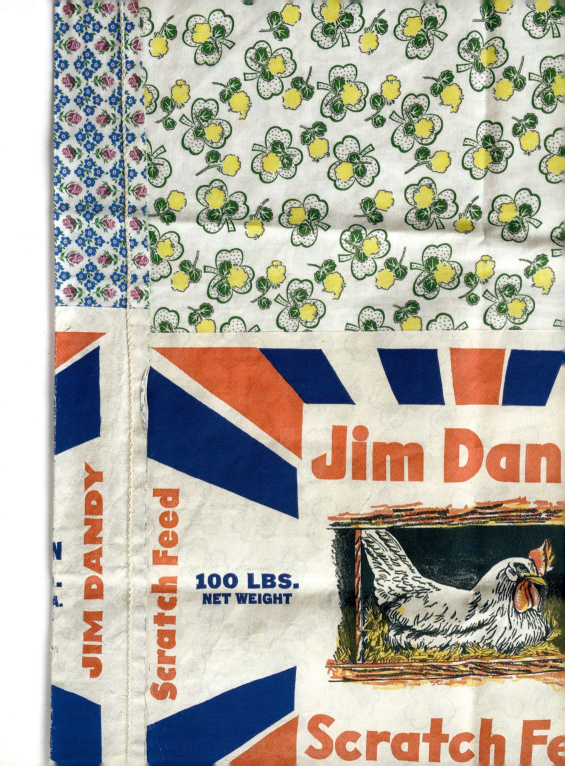

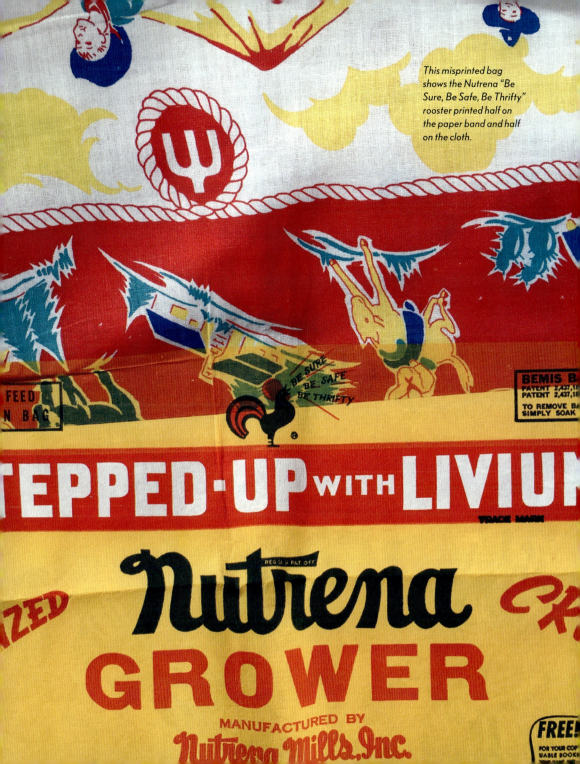

This misprinted bag shows the Nutrena "Be Sure, Be Safe, Be Thrifty" rooster printed half on the paper band and half on the cloth.

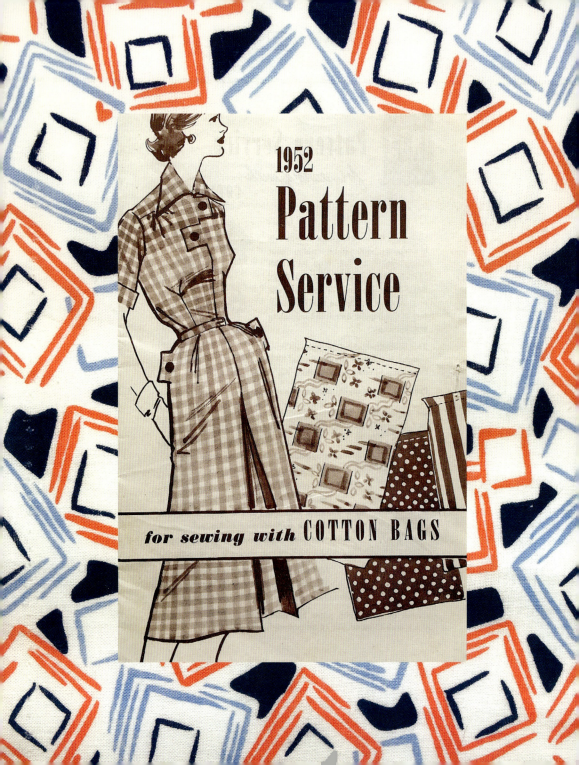

Pattern Service
for sewing with COTTON BAGS

Cotton bag fabrics, a long-time favorite with homemakers, will make fashion news throughout 1952! Your 1952 PATTERN SERVICE FOR SEWING WITH COTTON BAGS brings you scores of sewing tips and original ideas to help you take full advantage of this popular American homemaking practice.

This booklet, edited by the National Cotton Council, is a pattern service provided by your feed manufacturer and by the packers of other products which you purchase in thrifty cotton bags.

Probably you are already familiar with cotton bags and their re-use value.

If not, this booklet will introduce you to a completely new idea in home sewing. In any event we know you will welcome the fashion notes, illustrated patterns, and home decorating suggestions on the following pages.

We believe, too, that this booklet will bring you new thoughts about thrift. Though your husband has been buying farm commodities in cotton bags and you have been sewing with them for years, it's possible that neither of you has realized what a bargain you are getting.

Every time you buy feed, flour, fertilizer, or other products, you pay extra for the container. We call this "container cost." The manufacturer who packs his products in cotton bags saves you this expense. He gives you your money back — and *more* — in good quality sewing material.

A 100-lb. cotton bag gives you about one and a third yards of fabric that you can easily reclaim for sewing into dresses for yourself, clothing for the family, and furnishings for the home. A 50-lb. cotton bag gives you about one yard of thrifty sewing material.

And cotton bag fabrics this year are more beautiful than ever. There are smart plaids, florals, stripes, checks, modern designs, polka dots, and conversational prints adaptable to a host of sewing needs.

While prints are the topic of the day, you'll continue to find important uses for plain white and natural osnaburg bags. These take to dye readily, and can be used to carry out many clever ideas.

Fertilizer bags, too, will be a welcome source of sewing material in '52. Dress print fertilizer bags have joined the fashion parade. And do you realize that with a ton of fertilizer, cotton containers bring you more than 20 yards of re-usable fabric?

Cotton bag sewing is easy and fun for everyone. The Simplicity patterns in this booklet have been designed for special ease in converting your cotton bags into smart dresses. Consult your local sewing center for other ideas and helps.

Don't overlook the advantages cotton bags offer in 1952. Your sewing ingenuity will be the pride of your family and the talk of your community!

Cover dress —
a smart design in crisp cotton bag fabric. Simplicity 3704 (35¢), sizes 12-20, 40. Size 14 can be made from 3 bags, 40x46.

Smart and thrifty

Simplicity 3624 (35¢), sizes 11-18, a two-piecer for the junior miss. Size 12 takes 3 bags, 40x50.
Simplicity 3619 (35¢), sizes 12½-22½, frock with good lines. Add white collar and cuffs for a fashionable trim. Size 14½ takes 3 bags, 40x50.

Like mother, like daughter, *left*, a popular idea carried out in a cool and pretty design. Three bags, 40x46, will make mother's dress in size 14. Simplicity 3597 (25¢), sizes 12-20. Child's dress takes 1 bag, 40x54, in size 4. Simplicity 3598 (25¢), sizes 2-6.

A Pattern Layout with COTTON BAGS

Bags ripped, labels removed, and fabric neatly pressed... now you're ready for the pattern layout. Here's a good illustration of the way cotton bags can be adapted easily to standard patterns:

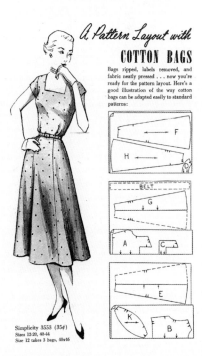

Simplicity 3553 (35¢)
Sizes 12-20, 40-44
Size 12 takes 3 bags, 40x46

PULLET POLLS PREDICT ELECTIONS

Modern political polls rely on quantities of data, plenty of cash and a little bit of luck, and even under the best of circumstances their predictions may fall wide of the mark. But in 1948, feed sack purchases predicted the results of the US presidential election more accurately than any professional pollster.

The election pitted Republican Thomas Dewey against the incumbent, Democrat Harry S. Truman. Truman's first term in office was his by default—as vice president, he succeeded Franklin D. Roosevelt, who died on April 12, 1945, just months after being elected for a fourth term. No one expected Truman to win re-election—a November 15, 1948, *Life* magazine story entitled "Truman Works a Political Miracle" described how his actions on election day demonstrated that he was sure he would be defeated. "Harry S. Truman, President of the U.S. by inheritance, slipped off to the Elms Hotel in Excelsior Springs, Mo. He had a mineral bath and a rubdown, ate a sandwich and drank a glass of buttermilk. Then at 7 p.m. he went to bed. It was a good time, if all the political portents as interpreted by the experts were right, to be alone and unconscious."

The experts who incorrectly predicted Truman's downfall included prestigious public opinion polls and news outlets. But one poll was recognized for being right—the "pullet poll" run by the Staley Milling Company of Kansas City. Earlier that year the company commissioned the Percy Kent Bag Company to create two labels for their bags of Pullet Atoms chicken feed. Bags with green labels read "Purchase of this sack is a 'vote' for the Democratic Candidate in the Staley Pullet Poll!" and showed a chicken astride a donkey. The red label that appeared on the other bags substituted the words "Republican Candidate" for "Democratic Candidate" and featured a chicken riding an elephant. The labels were affixed to bags sewn from a variety of prints.

The results of the pullet poll were tabulated and announced weekly. According to a story in the same issue of *Life* about the failure of traditional pollsters—"Press and Polls Were Wrong in a Loud Voice"—not everyone was off-base. "A few prophets had the truth in their grasp but refused to recognize it," they wrote, and noted that when the Staley Milling Company pullet poll showed that the Democrats had a 54 to 46 percent lead, "the company abandoned the poll in confusion." The *Sundance Times* from Wyoming printed in a November 11, 1948, article titled "Skeptics Backed Out of Pro-Truman Poll" that the pullet poll's results so differed from those of professional pollsters that "unbelieving company heads informed the distributors to stop the poll because it was 'not responsive.'"

When Staley Milling realized their poll had been accurate, they decided to give it another try. The *Kiowa News* from Kansas wrote on October 30, 1952, about a pullet poll conducted by J. E. Dainton, manager of the Bowersock Mill and Power Company, which sold Staley Milling products. The paper noted that "except for the week or two shortly following the Democratic convention Eisenhower has led in the weekly pullet polls as they were announced, and has led in the total at all times." The article ended, "The company and Mr. Dainton make no claims for the poll, but remember that it was right four years ago." Indeed, in 1952 it was right again—Eisenhower won.

Three Little Girls In Blue
Plus Hit Cartoon!

★★★★★★★★★★★★★★★

Calling All Republicans

in Staley's Pullet Poll

Come in today and cast your vote in the Staley PULLET POLL — the *only* poll that was right in 1948! And at the same time, hustle your pullets into egg production with PULLET ATOMS. Feed PULLET ATOMS 50-50 with grain.

FOSTER FEED STORE

701 Cherokee   Phone 123

No. 307-52   Staley Milling Co., Kansas City 16, Mo.

FOR YOUR BANK
ACCOUNT OR LOAN

Vote
FOR YOUR CANDIDATE IN THE
PULLET POLL

"THIS SACK IS A 'VOTE' FOR TRUMAN"

"THIS SACK IS A 'VOTE' FOR DEWEY"

Two different PULLET ATOMS sacks... one with the Democratic donkey; the other with the Republican elephant... give you an opportunity to back your candidate in the PULLET ATOMS Poll! Every purchase of PULLET ATOMS is a "vote" for one of the two candidates... we're keeping score, and we can tell you who's ahead in this vicinity.. in the state... in the entire middlewest! Come in and cast your "vote"... find out how the Pullet Poll is going!

FRED M. LANGE

308 W. Main St.   Sedalia, Mo.   Telephone 63

In The Bags!

KANSAS CITY, Nov. 8 (AP) —The "pullet poll"—a hen house test on White House preference—turned out to be right but its sponsor was so skeptical it was ashamed to announce the results.

Last July a milling company started a poll on the presidential race. It put its poultry feed in two kinds of sacks, one labeled "A Vote for the Republican candidate"; the other "A Vote for the Democratic candidate."

The first week Democratic sacks outsold Republican sacks by 8%. By Sept. 1, Truman had a 54% edge in the "pullet poll."

Thomas W. Staley, president of the company, took a look at the Gallup and Roper polls and decided his was falling flat.

"Sure," he said today when the "pullet poll" story came out, "we had more than 20,000 votes in by that time (Sept. 1), covering Missouri, Kansas, Iowa, Nebraska, Arkansas, Oklahoma and Illinois but we thought the figures were completely out of line."

Pullet Poll Was Right, But Pollster Wouldn't Believe It

KANSAS CITY, Mo. —(UP)— all the presidential polls went d-wrong for Dewey.

ut the pullet poll was right — a midwest pling from the heart of Amer- — tossed his aside because he n't believe his own reports. This is the sad story of Tom ley and his pullet poll.

t had the political dope aight from the feed-box as rly as the first week in September.

egun early in July as an advertising campaign for Staley's ling company and its chicken , the sampling scheme was ded a little differently than most famous national ear-

and listened to other advices. The big time pollsters—and the half a hundred Washington political writers who voted unanimously— said Dewey was "in."

Staley compared Pearson, Roper, Dr. Gallup with his figures — Truman 54 per cent, Dewey 46.

In early September, Staley decided his poll wasn't an accurate reflection of public opinion. He chopped off the pullet poll—the one that really was on the target of the true feeling of the American voters.

Feed Sack Poll Right Again

Staley Milling Company's "Pullet Poll," was right again in this presidential election. The pullets took a firm stand for Eisenhower weeks ago and predicted the Republican victory. It was the only poll wihch wts correct in the election of 1948.

The figures for the final week of the campaign were 54% Republican, 46% Democrat.

The Pullet Poll gave farmers a chance to register their "votes" by buying feed in either a Republican elephant or a Democratic donkey sack. Tom Staley, head of the Kansas City manufacturing firm, said the trend indicated that the farm vote strongly for Truman in 1948, had switched to the Republican side in 1952.

"Scientific or not, we're still the only poll with the record of never having laid an egg in a presidential election," Mr. Staley said.

Sale Dates

Closing out farm sale with

Room Refreshers

You can dress up every room in the house—at budget savings—when you sew with thrifty, colorful cotton bags. Make a new cover for that old chair, and repeat the color theme by framing squares of print for a bare space on the wall. Choose gay bags for breakfast room curtains, and add matching cloth or napkins for the table.

A pretty pillow will brighten any tired corner. Do covers yourself from distinctive bag prints and stuff with cotton batting... these are a happy choice as gifts for any occasion.

Simplicity 1394 (35¢), one size. Set of pillows with transfer included.

The pillow slip flour bag, available with bright borders or in quaint sprigged print, is a versatile favorite for home sewing. Ready-made for the linen shelf, it is equally useful for conversion into luncheon sets, curtains, chair covers, tea and guest towels.

MAKE THESE ATTRACTIVE AND PRACTICAL THINGS OF
Cotton Bags

Aprons
Smocks
House dresses
Beach coats
Combing jackets
Pajamas
Rompers
Children's aprons
Sunsuits
Middies
Mattress covers
Card table covers
Luncheon sets
Tray cloths
Jelly strainers
Handkerchiefs
Table runners
Garment covers
Laundry bags
Shoe cases
Broom covers
Pot cloths
Dish towels
Dusters
Bedspreads
Pillows
Curtains
Pillow cases
Ironing board covers
Vanity table drapes
Crib covers
Toast pockets
Bibs
Stuffed animals
Dolls
Refrigerator bags
Suitcase sets
Broom covers
Collar and cuff sets
Handbags
Yardstick holders
Book covers
Scrap books
Dress form linings
Hooked rugs
Muffin covers
Bean bags

Illustration and list of ideas from the booklet Sew with Cotton Bags.

HOW BAGS WERE USED

Reusing every bit of fabric was especially important to farm families, who often lived far from towns where fabric was sold and who frequently had little money to buy new textiles. There was a time in some rural households when every domestic textile—every diaper, dish towel, skirt and shirt—was made from a sack that originally contained feed, sugar, seed or flour. And these textiles were often recycled multiple times: clothing was handed down or remade for smaller family members, bits of fabric became patches or were stitched into quilts, and some of the most worn and damaged cloth found its final resting place in the rag bin, where it was used for cleaning, farm chores and animal bedding. The ingenuity and resourcefulness of rural residents was demonstrated through the many uses they found for feed sacks.

As fabric became more available to urban and rural residents alike, the National Cotton Council used advertising, press releases and "news" stories in small-town papers to encourage people to use feed sacks for diverse purposes. A 1946 article in the *Cullman Democrat* from Alabama acknowledged that feed sacks had mostly been used in utilitarian ways in the past, but proclaimed their more fashionable present and future: "this loyal family standby has moved out of the kitchen and into the bedroom and living room, and lives a glamorous life as a cotton fresh curtain, or new frock for milady."

SELECTING FEED SACKS

Many rural families spent their days isolated on their own farms—often only the man of the house went to town for items that could not be grown or made at home, while wives and children may have gone only once or twice a year. What had once been a straightforward process—going to the general store to buy bags of seed, feed, flour or fertilizer—became more complicated once feed sack manufacturers began printing sacks with embroidery patterns and dress prints. Stories abound of husbands sent to the feed store with a fabric swatch in hand to buy a bag that matched, or with explicit instructions to purchase the bag with the embroidery pattern that would complete a set of days-of-the-week dish towels. It also complicated life for shop owners, who were used to pulling bags from the top of a stack. Now it seemed the desired sack was always on the bottom of the pile, causing store owners to grumble as they shifted several 100-pound bags out of the way to retrieve it.

Nancy Freeland, who grew up on a dairy farm, remembers going with her father and three sisters to Gifford's Feed Store in Sharpsburg, Ohio, to choose sacks for her mother to make into skirts. The feed store visit was always exciting because they went just twice a year to buy feed to supplement the crops her father grew for his cows. "The bags would be all piled up and we'd climb and jump from stack to stack, looking for the one we wanted," she says. "I distinctly remember one print that had equations like 1 + 2 = 3, and I thought it might come in handy in school. You'd just have to look down at your skirt for an answer."

"When I was a little girl, we thought it was the biggest thrill in the world to get to visit local feed mills and pick out feed sack material for our dresses," recalls Mona Hicks in a September 11, 2002, article, "Thank the Lord for Feed Sacks," published in the *Rogersville Review*. "Mama always tried to go with us so she could pick and choose among

"Hey Maw, ain't you gonna be able to do anything with that feed sack?"

Cowpokes *was a regular cartoon about western and farm humour, started in 1952 by Ace Reid. It was published in hundreds of newspapers in the United States and Canada.*

the variety of prints and plaids available. Oh my goodness, there was an amazing variety. Mama always checked and made sure there were as few snags and holes as possible. You could buy feed sacks for just pennies a piece and sometimes they would give them to us. Of course, there were a few folks who considered it a shame to wear feed sack clothing. But we always considered it a blessing and I was proud to wear Mama's feed sack creations."

Diane MacLeod Shink worked at the store owned by her grandfather and father in the 1950s. Diane's great-grandfather had opened the store on the northern coast of Nova Scotia in 1880, at a time when there was no railroad in the area and all supplies arrived by boat. "It was a general store with a kazillion different things in it," says Diane. She remembers helping a woman find a feed sack that matched the one she had brought in with her. "She wanted to make a dress," Diane remembers. She led the woman to a warehouse behind the store where sacks of sugar, flour and animal feed were stacked, and they searched for a matching sack. She did not, however, think her father or grandfather would have taken the time to do this.

Danny Rogers grew up on a 200-acre farm in Garretts Bend, West Virginia. In 1959, when he was 16 and could drive, he was sent to town to the C. B. Cash and Sons general merchandise store. The shop carried everything a rural family might need—shoes, tools and groceries, as well as the feed Danny bought for the family's dairy cows. Danny remembers his mother's requests for thick-sliced garlic bologna, which was 50 cents a pound, and her instructions to always buy two matching feed sacks so that she could make a dress for his younger sister.

Suburban and urban women who wanted to sew with bags could get them from bakeries or farmer's markets, or from bag converters, who bought empty bags from bakeries and other places and resold them through mail order. Some grocery stores also sold flour, cornmeal, sugar, salt and other products in printed sacks. An August 9, 1948, article headlined "Dress from Paree? No—Just a Sack" in the *Democrat and Chronicle* from Rochester, New York, noted that farmers have long known that you could craft a fancy evening gown from a feed sack, "but we city folk have just begun to catch on—and feed and grain dealers in Rochester report a real 'run' on the feed-bag market."

A woman selects her favourite prints from a stack of feed sacks, New Mexico, 1951.

Save Money

MAKE THESE ATTRACTIVE AND PRACTICAL THINGS OF *Cotton Bags*

SIMPLICITY THE KEYNOTE TO TRIMMING

Elaborate trimming would be out of place on these things. A little simple and quickly worked embroidery is just as effective and far more in keeping. Outline stitch, lazy daisy and cross-stitch are among the best. Buttonholing is an alternative to binding the edges with the bias tape.

On the whole, embroidery in striking colors—red, green and royal blue—will work up much more effectively than the pastel shades. Another very satisfactory means of decoration is to cut out motifs from cretonnes and appliqué them on. A quarter of a yard of some juvenile pattern will trim several rompers, bibs and aprons.

DYEING

The woman who is handy with dyes will find this material takes the dye admirably. The colors are deep, solid and lasting. Some very clever tie-dyeing has also been done. In dyeing cotton, however, it should be borne in mind that the dyes must be stronger and be applied longer than with thin silks.

Aprons	Middies	Shoe cases	Ironing board covers	Collar and cuff sets
Smocks	Mattress covers	Broom covers	Vanity table drapes	Handbags
House dresses	Card table covers	Pot cloths	Crib covers	Yardstick holders
Beach coats	Luncheon sets	Dish towels	Toast pockets	Book covers
Combing jackets	Tray cloths	Dusters	Bibs	Scrap books
Pajamas	Jelly strainers	Bedspreads	Stuffed animals	Dress form linings
Blouses	Handkerchiefs	Pillows	Dolls	Hooked rugs
Rompers	Table runners	Comforter Covers	Refrigerator bags	Muffin covers
Children's aprons	Garment covers	Curtains	Suitcase sets	Bean bags
Sunsuits	Laundry bags	Pillow cases	Broom covers	Window Shades

And So—To Bed!

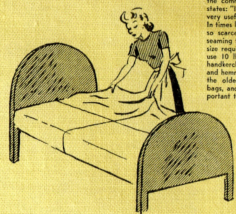

Even while feed bags are serving these many decorative and useful purposes, they are being turned also into sheets, pillow cases, mattress covers and dustcovers for bed springs. Typical is the comment of one reader who states: "I find that 'feed bags' are very useful in making bed sheets. In times like these when sheets are so scarce I use fine large bags, seaming them together to get the size required for each bed. I also use 10 lb. sugar bags for men's handkerchiefs, squaring the bag and hemming." These are some of the oldest uses for fine cotton bags, and there is none more important today.

Handy Andy Pieced Quilt

P-132

P132—Gives complete instructions for piecing and making this old fashioned "geometric" quilt—from scraps of feed bag materials and other cottons you have saved. Quilt takes 21 plain blocks and 21 pieced.

SHEETS AND BEDDING

This illustration from Bag Magic touts the utility of sacks for sheets, pillow cases, mattress covers and more.

The Kasco Journal suggests using leftover scraps to make the Handy Andy Pieced Quilt, 1942.

Border prints, as shown in the background of this page, were perfect for making pillow cases, curtains, and tablecloths.

Some feed sacks had instructions printed on them for making sheets: Percy Kent designed bags printed with embroidery patterns on one end that could be used as a "full size" pillowcase, but also included the instruction, "Sew 4 together for sheets." Ruth Rhoades interviewed rural Georgians who remembered sleeping on guano-sack sheets. One noted that if you had an itch, you did not need to scratch it: rubbing against the coarse sheets provided relief. (The fabric would eventually soften after many washings.) Cecilia Reid, keeper of her mother Pat's extensive feed sack collection, still sleeps on sheets made from sacks. "They're sugar sacks, actually, and I love them," she says, noting that they are more finely woven than animal feed sacks. "They're so nice and soft."

In small town America, news about bed linens created from feed sacks made headlines in local papers. Alongside articles on wildcat wells and factory inspections, the front page of the March 21, 1935, *Corrigan Press* from Texas included the story "Mrs. Peebles Makes

Mattress Cover." A member of the Lime Ridge Home Demonstration Club, Mrs. R. S. Peebles washed, bleached and pressed eight feed sacks to make a mattress cover "which can be removed and washed at any time." To avoid lumps, she used "flat felt [sic] seams." Mrs. Peebles' frugality was applauded: "Since buttons from discarded garments were used to fasten it at one end, the only cost of the well made, substantial cover was the thread used in making it." The article also noted that she next intended to save sacks to make a mattress pad, commenting: "Other members, inspired by Mrs. Peebles' example, are saving fertilizer sacks for the same purposes."

An article about a "Victory Sale and Exhibit" in the October 29, 1943, *Caldwell News and Burleson County Ledger* from Texas cited the display of the many items club members had been studying during the year, including canned food, upholstery, insect control, oil paintings, and bedspreads and quilts made of feed sacks.

Mrs. Peebles makes headlines on March 21, 1935, in the Corrigan Press from Texas.

The pamphlet Thrifty Thrills *encouraged decorating with feed sacks—and promised magic in the bedroom!*

Kitchens SHOULD BE MODERN

Windows can be made pleasingly attractive by using the Cotton Bag material in its natural shade, trimmed with a rick rack border of any contrasting color to harmonize with the kitchen's color scheme. Another idea is to sew the rick rack diagonally across the curtains each way in crossbars. The rick rack laid about 8 inches apart will give a pleasing diamond effect. Have at least two pairs for the kitchen so that you can put up clean ones at a moment's notice. Two bags will usually make a pair. Sewing brass or celluloid rings to the back of the hem at the top in place of a casing for the rod will make the curtains draw easier.

Trimmed tea towels nicely folded and conveniently placed add a touch of neatness to the room.

A simple luncheon set with napkins for the youngsters' noonday lunch materially reduces housework.

A stool cover to match your window curtains, which can be removed for easy laundering, adds a pleasing touch.

Washable Window Shades

To make window shades all you need do is take cotton bags, cut them the width required but be sure to allow enough for a very narrow hem on each side. The hem on the bottom of the shade through which you insert the stick can be made of any color desired. To add to its attractiveness it can be sewn onto the shade proper with rickrack—any color desired.

Use raw cold starch to stiffen the shade. Tack it on to a regular window shade roller, insert the stick in the hem at the bottom, fasten a colored pull cord on it and you have a novel, unique and lovely shade that can easily be laundered when soiled.

Hooked Rugs in Vogue

The popular hooked rug can be made as well with cotton bags as with wool, and costs less. Dye the bags (see page 31) without removing the stamping. Cut into strips one-third inch wide. It is not necessary to sew or tie the ends together. Use a coarse hooking needle.

«« 24 »»

A Dainty Vanity

A vanity table for the guest room or for your own boudoir makes a very attractive setting. It is practical and can be so trimmed to complete the color scheme of the room. Expert workmanship and expensive materials are not required. Any man handy with tools should be able to make a table with at least a shelf or a drawer or two.

A stool to match will be found practical and effective.

White curtains trimmed with colored ball fringe to match the general color scheme will complete the layout.

The vanity cover here pictured is made in three parts. The top is the last piece to fit. Cut your material to fit the table top allowing an inch margin on four sides to attach the ruffle. Use bias binding or rick rack on the ruffle. The same trim may be used on the skirt and overskirt. The main skirt should be thumb-tacked onto the edges of the vanity, pleating the material so that the skirt will be full. The overskirt should be trimmed to match the ruffle on the top and should be thumb-tacked over the main skirt. If you are handy with the needle designs may be appliqued on the overskirt.

If there is to be a shelf in the vanity table leave the drapery open in front with slight overlap.

An Ideal Material

In a cotton bag you have a strong sturdy piece of material, ranging in size from about 18 inches square to 36 x 42 inches, with two parallel selvage edges. It is firm, closely woven, and pre-shrunk. Muslin of equal quality, bought in a store, would cost considerably more. It will stand repeated tubbings and boilings and, since it cannot fade is exceptionally satisfactory for aprons and frocks, two articles of apparel which make frequent trips to the washtub.

A Novel Shower

How about a cotton bag shower? Get the bride's friends together, asking each one to bring a couple of cotton bags with her. Then go through this book and make out a list of your gifts and divide the work up among you. One or two afternoons of sewing and chatting over the tea cups will produce a small hope chest full of lovely useful gifts to present at the shower, that will please the prospective bride.

Sewing with Cotton Bags

Clothes Closets MUST BE MODERN

If you have been modernizing the rest of your home and neglecting the bedroom closets, here is your chance to bring them up to date at very little effort and less cost. Eleven flour bags will make you a complete set of three full-size garment covers, a four-pair shoe case and a personal laundry bag.

A Home for Shoes

First, the shoe case. The kind that hangs on the inside of the closet door is the cleanest and most convenient way of keeping shoes off the floor where they get dusty and scuffed. One bag, five yards of bias tape and two small brass or celluloid rings are all you need to make a case that would cost at least a dollar in any of the shops.

Cut off a strip 32 by 22 inches and fold in half so that you have a piece 16 inches wide by 22 inches long. This is the back of the case. The pockets are made out of two strips of the cloth, each 9 by 36 inches. The top edges should be bound with tape, which should be of some bright color, and the lower edge laid into four box pleats and then bound. The pockets are stitched down to the back, four inches apart. The outer edges of the case are then bound with tape all the way round. A narrow strip of the goods stitched about two inches from the top of the case holds the small strip of wood—a piece of old window blind slat will do—that keeps the case taut.

Sew one of the celluloid rings at each top corner and the case is finished. It will hold four pairs of shoes. By slipping out the strip of wood the case can be easily laundered.

Keep Out Dust

Then, the garment covers.

When you open your closet door do you see your best party frocks all jammed in with your everyday dresses, and dainty light voiles rubbing against heavy dark street dresses? If so, start today and make a set of garment covers for every closet in the house—three at least for each member of the family.

Flour bags afford excellent material for these covers because it is closely woven and will keep out the dust, is washable and inexpensive.

Three bags will make a full-size garment case. Cut the three pieces of cloth to a

width of twenty-seven inches. One of them, to be the front of the bag, is slashed up the middle to within six inches of the top. One slashed edge is faced back and the other has an extending facing to form a placket. Sew on snap fasteners at three-inch intervals. The top edges of the front and back are sloped down three inches to conform to the shape of the coat hanger, and are stitched, leaving an inch opening in the middle for the hook.

The third flour bag is folded in half along the twenty-seven inch width, and the long edges sewn to the bottom of the other two pieces. The sides are then stitched up and the bag is complete except for the addition of a small pocket stitched on the inside to hold a perfumed sachet or a few moth balls.

A Bag Apiece

A personal laundry bag for every member of the family is one of the marks of a well-ordered household and when one as attractive as this can be made for as little as ten cents, there is no reason for not having them.

A single sack, folded in half, is large enough for the ordinary laundry bag for personal use. Two probably will be needed if the bag is for household use, such as for sheets and tablecloths. This bag has several good features. The top is stitched closed, with a heading and a case for inserting a small rod or narrow slat of wood, such as are used for window blinds.

Soiled articles are put in through the slash in front and taken out by undoing the flap at the bottom. The word "Laundry" may be worked in cross-stitch or the personal element can be emphasized by working the owner's initials. The sides of the bag, the flap and the slashed opening are bound in colored bias tape.

TASTEFUL *Luncheon Sets*

Gone are the days when the housewife spent a solid hour ironing a white linen tablecloth into a state of glistening stiffness, or when a blob of gravy on the clean cloth was a domestic tragedy. For all except formal occasions the small family today sits down to a table laid with a tea cloth or with a centerpiece and individual place cloths, and napkins to match. The saving in labor is considerable as is the original cost, and nothing need be sacrificed in the way of attractive appearance.

For everyday use, these cloths are easily made out of cotton bags, which almost any baker will sell for a few cents apiece. Two or three such cloths will make a complete set, depending on the size of the table. Use oblong centerpieces and tray cloths for an oblong table and round ones for a round table.

Peasant Colors Effective

One effective style is embroidered in peasant colors, bright red and blue. The edges are buttonholed with blue and a simple cross-stitch design worked in each corner with the red and blue. A simple edging crocheted in cream-colored thread to match the cloth is another style. The border design is then worked in lazy daisy and outline stitch in pastel shades.

A plain hemstitched edge is a good finish, too, while for the breakfast room a colored border to match the color scheme is sufficient decoration with perhaps some simple insert of the color in the corners. The border for the tea cloth is about two inches wide, for the napkins half an inch.

Combined with Lace

Muffin and toast covers may be made to match the luncheon sets. Sideboard runners and tray cloths are treated in much the same fashion. A long sideboard runner is made by cutting one large cotton bag in half. One of the halves is cut in two again. These quarter sections are placed one on each end of the uncut half of the bag. Coarse lace is put all around the runner and inserted at the joins. This makes a runner two yards long.

«« 22 »»

FOR YOUR *Next Party*

Card tables have a variety of uses when their green baize or leatherette surfaces are hidden with attractive covers. Plenty of covers is one of the secrets of a successful bridge party. Not only must there be a clean, smooth one on each table for playing but the really fastidious hostess likes to put on fresh ones before refreshments are served. Again, when sewing on very delicate, perishable materials, as party frocks and dainty lingerie, the card table can be covered this way and used as a work table.

For Sunday suppers and for an informal tea or Sunday evening supper around the fire in the living room, nothing is simpler than to drag out the card table from its hall closet retreat, set it up, and cover it with a pretty cloth.

There is no reason why the thrifty housewife should not have a number of these always on hand since they can be very easily made at slight cost by using cotton bags. Cut the bag to the size of your table top. The corners are usually snipped off in order to make the cover fit more tightly. The edges are bound with bias tape in some bright color or may be hemmed and buttonholed. Plain white tape is used for the ties.

A Touch of Handwork

The cover may be decorated in a number of ways. One is to appliqué to it a three-inch heart, spade, club and diamond cut out of some solid color bits of cotton you may have in your scrap bag. Another is to cross-stitch some simple but effective design into each corner. Still another idea is to insert crocheted medallions, but embroidery and decoration should be confined to the corners of card-table covers as otherwise the cards are apt to catch in it.

If you want napkins to go with the cover another cotton bag will make nine twelve-inch ones which are buttonholed and decorated to match.

*A*ttention Bazaar Workers!

A novel and fast selling bazaar booth is one displaying only articles made out of cotton bags. These things cost little to make, are quickly worked, and sell rapidly because they are so practical and can be reasonably priced. Try it for your next Church or Club bazaar.

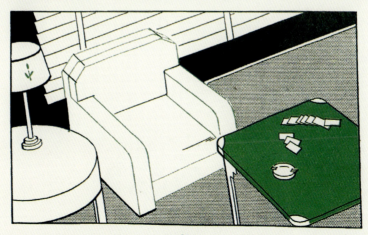

MAKE THESE ARTICLES FROM SMALL SUGAR, FLOUR AND MEAL BAGS

(See Size Chart Page 2)

QUILT PATCHES
Cut 13" square, take your favorite design and applique on bag goods. Assemble enough to make quilt 6 patches wide and 8 long. Bind patches together with tape.

BIBS
Bound with tape and tie string added a 5 lb. cotton bag can be converted into a child's bib and doll clothes. A little needle work added for design.

HOT PADS
Cut four 6" squares out of one bag, bind with bias tape, and add two rows of quilting stitches each way—a little brass ring may be sewed in corner.

DOLLS AND DOLL CLOTHES
Let the children make dolls and doll clothes. Stuffed animals are also welcome toys.

TRAY CLOTHS
A neatly embroidered tray cloth adds charm to your bread tray.

OTHER ITEMS AND USES
Refrigerator Bags
Doilies
Book Covers
Pot Holders
Bean Bags
Coffee Pot Bags
Jelly Strainers

Show your needlework skill on these items.

« 16 »

POSSIBILITIES FOR USING 10 TO 12 LB. BAGS

(See Size Chart Page 2)

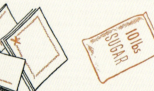

LUNCHEON NAPKINS
Dainty luncheon napkins with embroidered or appliqued design can be made from the 10 lb. bag.

PLAY HANDKERCHIEFS
Children are hard on handkerchiefs, especially when playing. Why not make up a supply from your sugar bags? They will last long and launder well.

COMBING CAPES
Just the thing to protect your clothes when making that 'last minute' hair dress or face powdering.

COAT HANGER CAPES
A good way to protect suits and dresses from dust is to make these practical shoulder protectors. They will pay for themselves many times by cleaning bill savings.

VEGETABLE BAG
An excellent way to keep vegetables fresh in the refrigerator. Add a draw string to the sugar bag and it's ready for use.

DUST CLOTH WITH MITTEN
Protect your hands from the grime of dusting by adding a mitten to your dusting cloth.

« 17 »

278

NOVEL SEWING IDEAS

Using 24 and 25 Pound COTTON BAGS

SEE SIZE CHART ON PAGE TWO

BABY'S LAYETTE

Here's an idea that's practical for the new baby. Any number of items may be made from the fine print cloth in cotton sugar bags.

KITCHEN CURTAINS

Most kitchen windows can be properly treated with half curtains. A little colored rick rack sewed on the borders will make attractive yet inexpensive curtains.

COVER FOR ROLLS

Keep your rolls hot by having a cover handy. A little embroidery will add to its attractiveness.

OTHER ITEMS

Napkins
Luncheon and Center Pieces
Pillow Slips
Table Linens
Covers to protect Pillow Tickings

Broom Covers
Tea Aprons
Tea Towels
Stuffed Toys
Dolls

Sun Suits
Dusters
Seat Covers
Laundry Bags

«« 21 »»

CLOTHES PIN APRON

A handy clothes pin apron can be made from 2 bags and a little binding tape.

HANDKERCHIEF CASE

Keep your hankies fresh by storing in this handy little case.

SUITCASE SET

Consists of a cover for frocks, a pair of shoe cases, glove and wash cloth case—line wash cloth case with rubber lining. Trim with contrasting binding tape.

«« 20 »»

IN THE KITCHEN

Growing and cooking food occupied countless hours in women's lives, and cotton bags had multiple uses in the kitchen. Women used coarsely woven sacks as strainers and sieves to help separate solids from liquids when making molasses or jelly, or when transferring lard to containers during butchering. Dish towels are one of the classic reuses of sacks, particularly flour sacks. Dressed up with embroidery, rickrack and bias tape, fancy dish towels helped enliven tedious tasks. So renowned were (and are) flour sacks for drying dishes that women who did not have access to them at home would buy them from department stores and bakeries. The *Canadian Advertiser* from Texas included an ad in its September 9, 1938, edition for Studer's Market and Bakery listing rich layer cakes for 25 cents each, Allsweet Oleo at 19 cents a pound and "Large Laundered Flour Sacks" for a dollar a dozen.

Home bread bakers used feed sacks to cover rising dough and keep it warm. Iowan Florence Stockman remembers that in the 1940s, her mother used feed sacks to protect plenty of other kinds of food, too. "We didn't have waxed paper or plastic wrap in those days, so we covered rolls and anything that sat out that you didn't want the flies to get to," she says. Her grandmother, who had a farm near Audubon, Iowa, relied on sacks to shield her baked goods. "She fed hired men and threshers, and there would be lots of angel food cakes that needed covering," says Florence. The men sat at card tables that her grandmother covered with feed sack prints, their edges neatly frayed rather than hemmed. Feed sacks were a valuable commodity, and though there would be the occasional hole in a sack where a mouse had chewed through, this did not deter these thrifty homemakers. "It was good cloth, and we certainly weren't going to throw it out," says Florence.

Illustration from Pattern Service for Sewing with Cotton Bags, *1952.*

One of the most common items stitched from feed sacks was aprons. Most could be sewn from a single sack. Dot Bloom of Rye, New Hampshire, recalls their popularity: "Everyone wore aprons, because in those days you wore a house dress til noon. Then you'd put on your good dress to go out in the afternoon and you'd put an apron over that when you came home to make dinner." (Apparently greeting your hungry spouse while wearing a house dress was not an option.)

An article in the *Abilene Reporter-News* from 1953 suggested that feed sacks were the perfect material from which to make Christmas crafts. In addition to "bedroom slippers, lamp shades, and dust mitts," the reporter recommended making ponchos "to take the place of aprons on your Christmas gift list this year." It was not clear whether the writer meant that gift recipients would wear ponchos in the kitchen or simply that they would delight in receiving ponchos in place of traditional aprons that year. Either way, the material of choice was feed sacks.

A collection of various potholders sewn from feed sacks.

GLORIA HALL COLLECTION

Presenting the cotton bag poncho

Cotton bags in 100-lb. sizes are made-to-order for ponchos, and you won't even need a pattern. We suggest that you wear these comfortable cover-ups around the house for working, playing, and entertaining.

To make a Cotton Bag Poncho:
Cut a 100-lb. bag into two lengths. Hem or bind edges. Sew shoulder seams at top. Add ties in the middle. Take your choice of necklines, straight, round, a V with lapels. Match a belt or sash to the lapels. Add your own decorative touches such as pockets, rickrack trimming, appliqued designs, or embroidery.

1954 Idea Book for Sewing with Cotton Bags

cotton bag aprons

This one-yard apron, in any one of four styles, can be made from a 100-lb. cotton bag. Simplicity 4443 (35¢) gives you a whole wardrobe of apron fashions, ranging from workaday to festive.

1. Both pretty and practical. Note scallops and trimming interest.

2. Sensible to work in. Wide pockets are well-placed and handy.

3. Strictly for show, with ruffles added to the scallops.

4. For the hostess. Becoming, in both living room and kitchen.

APRON THREESOME

The homemaker tied to her own apron strings won't shirk kitchen duty when her aprons are fashion-wise. Cotton bags provide fabric with style, and here are three smart, easy-to-use patterns. You'll have fun making a new apron wardrobe.

Left, Simplicity 3550 (25¢), sizes 12 to 20, 40 to 44, offers a unique dress and apron combination, trimmed in matching and contrasting fabric. Size 16 takes 5 bags, 40x54, for both. *Center,* a coverall apron with two deep, handy pockets. Three bags, 40x46, are required for size 14. Simplicity 3717 (25¢), sizes 12-20, 40-44. *Right,* an apron frock trimmed with binding. Three bags, 40x46, will make size 14. Simplicity 3745 (35¢), sizes 12 to 20.

Pattern Service

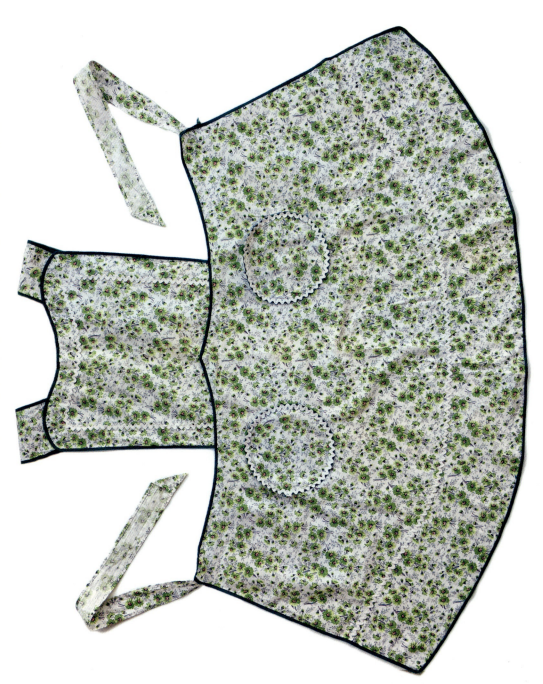

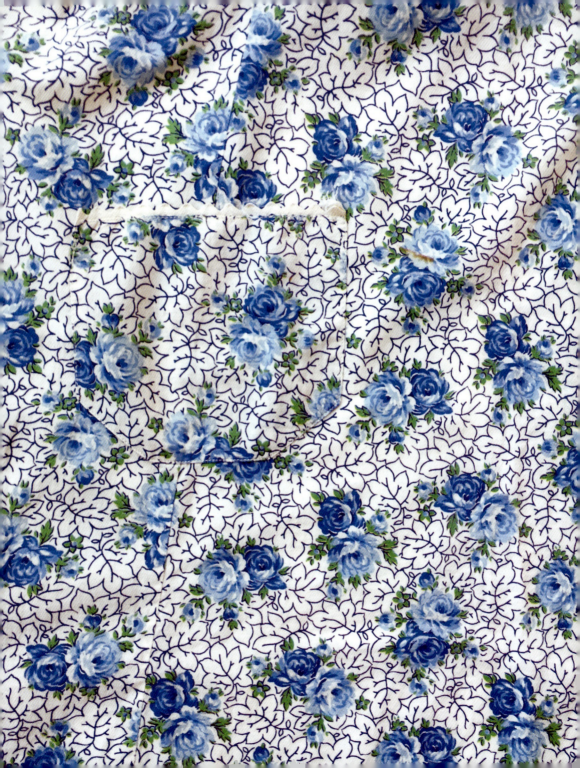

MAGAZINE PAGE FOR EVERYBODY

HOME • HEALTH • FEATURES • BEAUTY • FASHION • FICTION

Give Your Venetian Blinds Care

By ELEANOR ROSS

VENETIAN blinds, like everything else in the house, need care, and respond quickly to a good going-over.

Give them a quick dusting with an ostrich feather duster or a quick flicking-over, weekly with your regular dust cloths or vacuum attachment. This will keep the blinds in tip top condition. Always dust away from the tapes.

Once every two months or so, wipe down the inside of the blinds with any of the new standard cleaners or even plain soap and water.

Cleaners for the Blinds

If you are gadget minded, there are several venetian blind cleaners on the market, though none of them are any more effective than elbow grease.

To keep blinds in proper condition, functional and an ornament to the home, do not use wax on wood blinds. No matter how sparingly it is used, the wax makes re-painting of the blinds impossible. And wax does not make the dusting of blinds unnecessary, as so many women seem to think.

Do not douse your blinds in the bathtub! Tapes may shrink or fade, carelessly undried water spots may mar the beauty of the blinds, be they wood or otherwise. After all, one does not dunk one's other furniture or accessories in the bathtub—and blinds are furniture!

New Drug Just Discovered

By HERMAN N. BUNDESEN, M.D.

THE search for new drugs is endless. Medical chemists are constantly attempting to devise new and more effective remedies for every kind of illness. Recently, a drug known as dibutoline, which seems to have a powerful, soothing effect on smooth muscle tissue of the type which makes up the stomach and bowel, has been discovered. There are various disorders in which muscles of the intestinal tract tighten up excessively and go into spasm, thus causing periodic attacks of pain. One type of muscle spasm occurs in the condition known as spastic colitis, in which attacks of diarrhea alternate with constipation and there is marked pain.

When dibutoline was given to a group of patients suffering from spastic colitis, complete relief of pain occurred in one to 10 minutes. The relief lasted in some cases for several weeks and, in others, for from two to three hours. No other treatment was employed while the dibutoline was used.

Pain in Abdomen

There also is a disorder known as diverticulitis, in which pouches form on the large intestine. In this disorder also, pain in the abdomen occurs. Patients with diverticulitis, treated with dibutoline, were kept free from pain. The drug, however, had to be administered from three to eight times daily.

Dibutoline was also found useful in the treatment of ulcerative colitis, in which there is inflammation of the bowel, together with the formation of ulcers. The drug assisted in controlling the abdominal discomfort and lessening of the frequency of bowel movements.

Spasm of Muscle

The muscle between the stomach and bowel is known as the pyloric muscle. Spasm of this muscle may occur when there is an ulcer of the first part of the bowel, called a duodenal ulcer. When a spasm of this type occurs, pain develops. Complete relief of this type of pain was also produced with the dibutoline.

It would appear that this preparation is effective in relieving pains produced by spasm of the bowel muscle. The drug must be given by injection under the skin. So far, dibutoline has been used in relatively few cases, but when it becomes available for general use, it should prove a great value in a variety of intestinal disorders.

By MARION CLYDE McCARROLL

IT'S in the bag!

In the big flour bag; the sugar, salt, meal or feed bag.

What is? Why, a new cotton dress for the summer! Or a pair of curtains for the kitchen, a frilly skirt for the dressing table, slipcovers for the living room chairs.

Quietly making its way throughout the country — and already a common custom in many sections — is the trend toward making simple dresses and decorative accessories for the home out of cotton bags.

Farmers buy bags of feed, and when the feed is emptied out, their wives carry the empty bags into the sewing room. Women go to buy baked goods, and at the same time purchase for a few cents some of the 100-lb. bags in which flour is delivered to bakeries. They buy family-size bags of salt, flour and sugar at the super-market.

Then they rip the stitching off the bags, wash out the printed labels, launder the bags thoroughly, and there they have cotton yard goods ready to be made up into house coats, dresses, smart little jacket-and-dress costumes for summer, bedspreads, quilts, luncheon sets and a host of other good looking things at a cost of a couple of dollars or so.

Some of the bags come already printed in colorful patterns, others in pastel shades, stripes or plaids, while the white or natural-colored ones can be dyed as desired.

About four bags of the 100-lb. size, each slightly more than a yard long, are required to make an average dress, while children's clothes, of course, take fewer.

Shown here is the young wife of one of the G.I. students at a southern state university making and modelling a flour bag costume.

Anyone who would like to try their hand at cotton bag dressmaking can obtain full information about where the bags can be bought by writing to the National Cotton Council, P.O. Box 76, Memphis, Tenn. The Council also has an illustrated style booklet showing many attractive garments that can be made from them.

"I'LL TAKE A LOAF of bread, six cup cakes, and four cotton flour bags in a red-white-and-green plaid pattern," says Ann Pilcher, at the bakery.

BACK HOME, SHE QUICKLY ripped up the bags—an easy matter as they're chain-stitched in cotton crochet thread—washed and hung them out to dry.

Taking Your Beauty Props on a Trip

By HELEN FOLLETT

GOING places? Naturally! What do you expect to do with your beauty props? Time was when they were tucked into odd corners of the suitcase. Maybe the powder spilled out, and was that a mess? Sometimes the stopper came loose from the precious perfume. Over that grief you shed bitter tears. Nothing like that these days. Cosmeticians think of everything. And the traveler's good-looks equipment is something to sing thanks about.

When you are prowling around the shops, look them over. No use hanging tight to your money; you'll spend it and be happy ever after. Better to invest cash of the realm for your own personal loveliness than to have it in the bank where it isn't doing you a bit of good.

Small Size

The nice things about these kits is that every item is of tabloid size, just enough to do you for a month or two, then you can get refills. There are creams, a powder, rouge and lip stick, the Big Four in the cosmetic family. Sometimes there is space for eyebrow crayon, tweezers, manicure scissors, orange wood stick. Some come in wallet form with fewer cosmetics but space for folding money and spending change. Always it is nice to pack light when vacationing.

The girl who goes sight seeing will have to include a foot powder. The little dogs play out when one tramps through museums and art galleries.

Pretty pouches of satin, lined with rubber, are convenient for soap and washcloth. One is always in a quandary about what to do with the moist cloth and most of us feel that we should carry it, no matter if the cleansing cream is brought into action, too. And there's the bath brush. That goes in with the soap and the cloth. Once used to a brush, no bath seems complete or thorough without it.

Smart girls keep a list of toiletries pinned to the lining of the suitcase. Then no item is forgotten.

LAYING OUT HER PATTERN on the clean bag pieces, she first cut out the wide skirt; then the other parts of what was to be a dress-and-jacket costume.

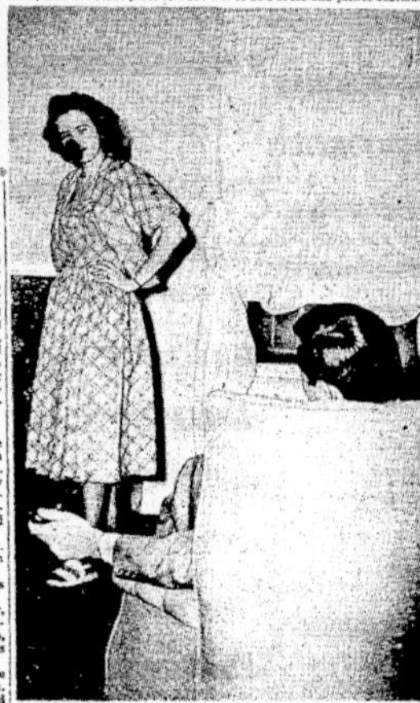

WHEN HUSBAND BOB GOT home that night, she was ready to model her smart new summer outfit which cost her well under three dollars.

Two Kinds of Flour

* * *

Old-Time Kitchens Included Them

By IDA BAILEY ALLEN

"WHAT is this kind of flour, Madame, that is called G-R-A-H-A-M," inquired the Chef, looking up from an old cook book he was studying.

"It's a form of entire wheat flour, invented many years ago by an American doctor, Sylvester Graham, who was one of the first food specialists to recognize the value of whole grains as food. In both my grandmother's and mother's kitchens, there were always two kinds of flour, 'white' or all-purpose, and 'graham' flour."

"I have not seen it on sale anywhere," said the Chef. "Can you buy it today?"

"Very rarely under the name 'graham flour'. But another form, milled, so the flour is a little less coarse, can be bought everywhere. It's called entire wheat, or sometimes whole wheat flour."

Whole Grain

"And is this really made of the whole grain, Madame?"

"Yes, with this exception—the flinty outside covering of the wheat is milled off. But the most important part, which is the heart of the wheat and contains the vitamins, is not removed. Entire wheat flour is one of our fine body builders, and also supplies bulk necessary to help along intestinal action."

"I notice in the recipes in this old cook book, Madame, that the graham flour seems to be combined with an equal amount of white flour, and that it is used in making yeast bread and rolls as well as pancakes, muffins, the tea biscuits and fritters. Why was it necessary to use so much white flour?"

"That's because when the outer coats of the wheat, called bran, are ground into the flour, there is naturally less starch and gluten in each cupful. So the coarser the entire wheat flour, the more white flour must be added to make a dough elastic enough to be raised, either by yeast or baking powder. Modern millers felt the flour would be even more healthful if it were milled a little finer, and fewer husks were included. So today much of the entire wheat flour we buy is milled fine enough to be sifted, and can be used instead of white flour in any recipe for quick breads. But if it is to be made into yeast bread or rolls, and a 'light' bread is desired, about a third of the flour used should be white flour. If the loaf is made wholly with entire wheat flour, it will not rise so high, and will be a little heavy and coarse in texture, but still appetizing."

Good for Health

"It would be good for the health to use more entire wheat bread, Madame."

"Yes, it would certainly help to raise the standard of good health, Chef. But most families hate to change their eating habits. They're accustomed to white bread; so I'd suggest providing an equal number of white and entire wheat loaves of bread each week. Vary the breads from day to day. Then if our homemakers will buy a small bag of coarse whole wheat flour and use it once a week to make entire wheat muffins, biscuits or pancakes, their families will become accustomed to the unfamiliar texture, and soon ask for 'more'."

TOMORROW'S DINNER

Thin Cream of Corn Soup
Croutons
Medallions of Beef with Onions
Hominy Grits Spinach Loaf
"Graham" Muffins
Strawberry Sponge
Coffee or Tea Milk (Children)

All Measurements Are Level
Recipes Serve Four

Thin Cream of Corn Soup

Melt 2 tbsp. butter or margarine in a qt.-sized sauce pan. Stir in 1¼ tbsp. flour, 1 tsp. salt, ¼ tsp. sugar and ¼ tsp. white pepper. Gradually add 2 c. milk, either fresh fluid milk or reconstituted dry skim milk. Add ⅓ tsp. scraped onion juice, ½ (No. 2) can cream style corn of good quality and 1 c. boiling water. Place in a double boiler and cook 20 min. over hot water. Rub through a sieve if you like. Personally, the Chef and I prefer the corn served in the soup.

This may be served hot or well chilled. Dust each serving with fine-minced parsley or chives for the gourmet touch.

Medallions of Beef with Onions

Order 1 lb. sliced top round steak cut ¼" thick. Cut it into 6 neat pieces, then pound with a meat mallet until flat and oval in shape. Mix together ½ tsp. salt, ¼ tsp. pepper and 4 tbsp. flour; rub into the meat. In a large skillet melt 3 tbsp. butter or margarine. Fry the steak in this until brown on both sides. Pour in ½ c. boiling water and add ½ tsp. meat extract. Cover and simmer 40 min. Then cover with steam-fried onions and slow-cook 12 min. longer. Serve with hominy grits.

"Graham" Muffins

Mix together 1 c. coarse entire wheat flour or "graham meal", 1 c. all-purpose flour, 2 tbsp. sugar, 3 tsp. baking powder and 1 tsp. salt. Stir in 1 c. milk, 1 well-beaten egg and 3 tbsp. melted butter, margarine or shortening. Transfer to good-sized heated oiled muffin pans, filling them half full. Bake in a moderate oven, 375-400 F. Makes 8 muffins.

Strawberry Sponge

Add 1 envelope unflavored gelatin to ¼ c. cold water. Stand 5 min. to soften. Then add ½ c. boiling water and set over steam until dissolved. Add ¾ c. sugar and when cooled, 1¼ c. mashed strawberries and the juice, and 2 tbsp. lemon juice. Chill in the refrigerator until the consistency of honey. Then add 2 egg whites beaten stiff, and beat steadily with an egg beater or an electric mixer until very fluffy. Transfer to a mold first dipped in cold water. Chill; unmold and serve garnished with sweetened whipped cream or dry skim milk topping, and halved whole strawberries first rolled in granulated sugar.

TRICK OF THE CHEF

To give a new taste to spinach, crush ½ section peeled garlic and fry a minute in 1 tbsp. butter in the kettle in which it is to be cooked.

Food for Thought

If, in spite of all your care, you have stale bread, use it in meat loaf, meat cakes, cheese fondue, brown Betty, griddle cakes, stuffings for poultry, tomatoes, or green peppers, in custard bread pudding, etc.

In buying sweetbreads, allow one pair for each two people to be served.

LOVE'S PERILOUS PATH

A Sequel to Love's Fair Horizon

By ADELE GARRISON

* * *

As Dicky Finishes His Talk With Faith, the Boys Jump Up, Saying They Hear the Front Door Knocker

Synopsis: In the Hotel Lausfold suite of her father, Dicky, little Faith Graham admits she was snooping when she overheard her parents talking about the "brainstorm" of Noel Veritzen, husband of Dicky's niece, Mary. The child does not cry because she has promised not to during Dicky's interrogation following her indiscreet repetition of this phrase, but now, suddenly upon the verge of breakdown, she begs for a switching instead of continued questioning. "There'll be neither," Dicky says tenderly, "but I want you to make me a promise." "You want me to promise not to snoop?" Faith queries terrifiedly. "I can promise to try not to, but if I broke it, I'd be a miserable sinner."

I COULD not see Dicky's full face, but a look at his profile told me that his lips decidedly were twitching, as they generally do at his small daughter's choice of odd words.

"That isn't the promise I want you to make," he said, "although I should like very much to hear that you will try not to listen at doors any more. But I do want you to promise me very solemnly that never again will you repeat anything you have heard in your eavesdropping when anybody except your mother or me can hear you. You see you could have made Cousin Mary very sad, if she had heard you say that about Cousin Noel, for she would have known that your mother or I had said it first."

... ing determination to save her husband from the thrall of his temper fits, first by insisting that the tantrums were not hereditary, and then being equally firm in her decision not to bring a child into the world with such a heritage, if his father's theory were true, had brought Noel to his senses. Until his recent explosion about Georges, for which, I admitted he had had ample provocation, Noel had shown no symptoms of tantrums since long before Mary had left for the sanitarium, and none since she had returned.

Many Memories

But I had to thrust aside the memories which had rushed with the speed of lightning into my mind. Faith was speaking, and, not to my surprise, for her resilience is astonishing, her words were the flippant, slangy ones she has heard ...

This article from 1949 chronicles Ann Pilcher making a dress from sacks. "When husband Bob got home that night, she was ready to model her smart new summer outfit which cost her well under three dollars."

... trums, and those of Noel, were a royal heritage, even though "bar sinister," of the ruling house of Transvania, and that there was no known cure for them.

Mary's common sense and her lov- ...

... upward, outward stroke until dry.

In making up the lips, follow the natural outline as closely as possible. Don't enlarge the outline unless so advised by a make-up expert.

fashion in rhyme and reason

By *Elisa Mattley*

Down to the sea
In scarf and shorts,
Perfect for swimming
Add a shirt for sports

6-B

How Parents Can Help To Develop Child's Imagination

By GARRY CLEVELAND MYERS, Ph.D.

IT'S wonderful how some appreciative mothers sense the gradual development of imagination and creative powers in the young child and encourage this child to go on creating. For example, one mother writes:

"Next door, Madge has the mumps. Our Danny who is nearly five takes a letter to her each day. I write the letter which he helps me compose. Since he was playing with Madge the same day she actually became ill, we expect him to get the same soon. Meanwhile, he and I seek to brighten Madge's days in bed. He and I have great fun doing it."

Sample Letter

She enclosed a sample letter she had written to Madge from the suggestions and words of her little boy. In this way she set a strong motive for her son to create from words, for fun with a purpose to make Madge happier. Can you think of finer furtherance of unselfish idealism while making two young children happy? Besides, the mother gains a lot of satisfaction for herself.

Another mother writes: "My daughter takes great interest in the page of printed drawings in a children's magazine she has. She has always been exceptional in art and I have filed her drawings, dating each one. Knowing of your deep interest in and appreciation of the drawings of young children I thought you might like to see the three large scrapbooks I have made of Milicent's creations. As they are precious to her and me I wish you would return them to us."

First Volume

They are truly wonderful. I wish you could also see them. The first volume is labeled, "Milicent Ann Drury's Drawings at Four Years to Kindergarten Years." The second, "At Six Years of Age." The third, "At Eight Years of Age."

Let us hope that Milicent is now making her own scrapbook, her mother helping her, enjoying it with her. Of course, the scrapbooks will be returned. Milicent and her mother will enjoy going through these books to observe the young artist's growth. Years hence, Milicent's grandchildren may also enjoy these books together with their own scrapbooks.

As one goes through these books one not only enjoys the creative achievements and growth of Milicent but also the appreciative wisdom and foresight of her mother, without whom these books would not have been, nor Milicent so far advanced as a creative artist.

I wrote to the mother and Milicent asking their permission to use their names and addresses in this column, in order to make this report more real and interesting to you. Gladly they granted it. Mrs. Thomas W. Drury and her daughter Milicent live at Sturges, Michigan, P. O. Box 42. Let me add that while anything I quote in this column is from a real person, I never use a name or address without permission. Nor do I ever quote from a letter which is marked "confidential."

I wish thousands of my readers would emulate Mrs. Drury. They might also record the yarns the little child makes up and make a scrapbook of them, not only to stimulate the child to create more but to have wholesome fun and companionship with him. Related bulletins of mine are: "Now Come The Fairies," and "Letting Little Children Learn," to be had in a stamped envelope sent me in care of this paper.

THE STARS SAY—

By GENEVIEVE KEMBLE

For Thursday, June 9

WHILE there is augur, of a slowing down of lively conditions, with some block, limitation or postponement, yet the situation may not prove entirely hopeless or arrest the wheels of progress already established on a promising basis, with reasonable and logical foundation for growth, expansion and financial stability. Such encouraging situation could be easily muddled or disintegrated by some reckless, forced or militant action, alienating the expected support. Maintain calm and await normal developments.

Those whose birthday it is, may find themselves well established on a firm and expanding basis, with promise of aid from influential sources, although there may be flashes of disappointment, postponement or other frustration or limitation. This should be philosophically evaluated, with sound analysis of enduring factors and underlying stability. Any attempts to force the issue, to batter down opposition, or show other forms of impatience, bad temper or impetuosity, could bring about complications as well as personal danger. Keep calm and work for long-range results.

A child born on this day may be hot-headed, impetuous and disposed to force and fury, although its future prospects and enduring returns could grow by cool management.

Words of the Wise

Beware of him that is slow to anger; anger, when it is long in coming, is the stronger when it comes, and the longer kept.
—(Francis Quarles)

Marriage is the one subject on which all women agree and all men disagree. —(Wilde)

MAKE THESE *Youthful* FROCKS WITH FEED BAG FABRICS

8968—The peasant frock is a simple flattering style with the new full skirt. 12 to 20. Size 18 takes 4 bags, 39x43 inch size. Fold skirt portions. Piece at side.

8929—Here is a pretty neckline and new, open sleeves. Snug waist line, too. 12 to 20. Size 18 takes 3½ bags 39x43 inch size.

#967—Smooth midriff and swingy skirt as well as a "frosting white" collar are smart details in this frock. 11 to 19. Size 13 takes 3½ bags. Collar takes ¾ yard of 35-inch contrast material.

A POPULAR NURSERY TOY

7758. A stuffed goose that may be made to look attractively real. It may be a favorite toy, or a comfortable cushion if filled with down. Designed in One Size: 15 inches in length. It requires 3 small flour bags 16 x 18, or one large bag, and a piece of gingham or contrasting material, ½ yard long and 7 inches wide for the bonnet. The tie strings of ribbon require ½ yard. Price 10c.

FUN AND GAMES

Although farm and small town life involved plenty of backbreaking work, there was still time for play. Phyllis Rosenwinkel recalls that in the 1950s her father made her a swing by stuffing a feed sack with hay and using a long rope to tie it to a branch of a maple tree in a grove behind the family's farmhouse near Clarksville, Iowa. Phyllis envied a friend who had a hula skirt made of raffia, so she fashioned her own from feed sacks that her mother dyed red. Phyllis cut numerous slits in the bottom of the sack and then ran a string through holes in the top to gather it around her waist. Later, Phyllis and some friends wore feed sack hula skirts as part of a circus put on by her Brownie troop to entertain their parents. She also stuffed a feed sack with hay for insulation and used it to prevent a "cold butt" when her dog pulled her along on her metal saucer sled in winter. "This kind of thriftiness wasn't a virtue," says Phyllis of her family's recycling and reuse. "It was just the normal way of living."

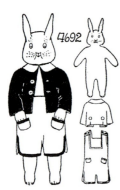

"PETER RABBIT" AND HIS WINTER SUIT

4692. Here is a well known nursery friend, with a new Jacket and Overalls. The Pattern includes the "doll" and the garments. It is cut in 3 Sizes: Small 12, Medium 16, Large 20 inches in length. A 12 inch size requires for the doll and garments, 2 small 26 x 26 flour bags. If Peter Rabbit's jacket is made of contrast—it will require ⅜ yard 35 inches wide. Price 10c.

Doll clothes and quilts for dolls were frequently made from leftover bits of fabric. There are many reports of dolls and their owners wearing matching dresses, the larger dress stitched from full feed sacks and the smaller stitched from the scraps that remained. Children enjoyed the cut-and-sew dolls printed on the back of sugar sacks, and simple stuffed animals, sewn from scraps, were popular as well.

While feed sacks were part of everyday life, they also showed up on special occasions. Feed sacks served as the basis of many Halloween costumes, from the most basic ghost—holes cut in a white sack for arms and eyes—to more elaborate outfits. Gold Medal Flour offered customers a pattern for a three-flour-sack clown costume in the 1920s.

Ruth Rhoades notes that in Dawsonville, Georgia, during World War II, high school graduates contributed to the war effort by foregoing costly caps and gowns, instead making their own graduation dresses out of feed sacks. The floor-length gowns were individually sewn, and no two were alike. Two of the graduates were so proud of their creations that they still had swatches of their dresses saved when Rhoades interviewed them in 1997.

Loris Connolly writes that the Chase Bag Company's bowling team wore shirts stitched from feed sacks, while a bakery-sponsored bowling team wore flour sack shirts.

While feed sack creations were objects of pride for some, for others they epitomized a shabbier mien. A story in the October 19, 1948, *Fort Worth Rambler* from Texas noted that as part of their initiation rites, pledges to Illotus Duodecim, a fraternal organization at Texas Wesleyan College, were made to wear "feed sacks as shirts [and] pants turned backward." (Further acts of degradation included having their hair plastered with axle grease and being required to carry a dead fish at all times, "except in chapel and classrooms.")

"Grandma" humour from 1955 has an old woman playing football and putting a new spin on the notion of "getting sacked."

These 1950s craft projects for boys include an indoor punching bag and a burlap "sad sack" figure.

There's Fun in Doing Things TOGETHER

Going to Town

Shopping Together

Baking with Mother

Sewing Circle

Family Picnic

DURING the formative years of a girl's life, there is no single thing which molds her character so much as doing things *with* and *for* mother. Pyschologists count this the most important factor in building a wholesome approach to living. They need not be important things, though these count, but the dozen and one little daily tasks which fitted together make up the patchwork called daily living.

Cotton bags can be a factor encouraging this close companionship. Just as men of the family collect stamps, the womenfolk can collect bags and plan ways to make home bright and living gayer through them. Once this collecting is begun, it is natural to ask, first, for goods packaged in cotton bags.

The ripping and preparing of the bag, the sewing circle which mother and daughter begin results in new and attractive clothes, in pretty accessories around the home. Talents so modestly encouraged develop easily, naturally.

Recommended for mother and daughter projects are not only such easy-to-make accessories as picnic cloths and napkins, home decorations as shown throughout this booklet, but the mother-and-daughter fashions on the opposite page. It is difficult to say who is more pleased with these — mother who wants to stay young in heart, or growing daughter who wants to copy quickly everything that mother wears.

A Bag of Tricks for Home Sewing

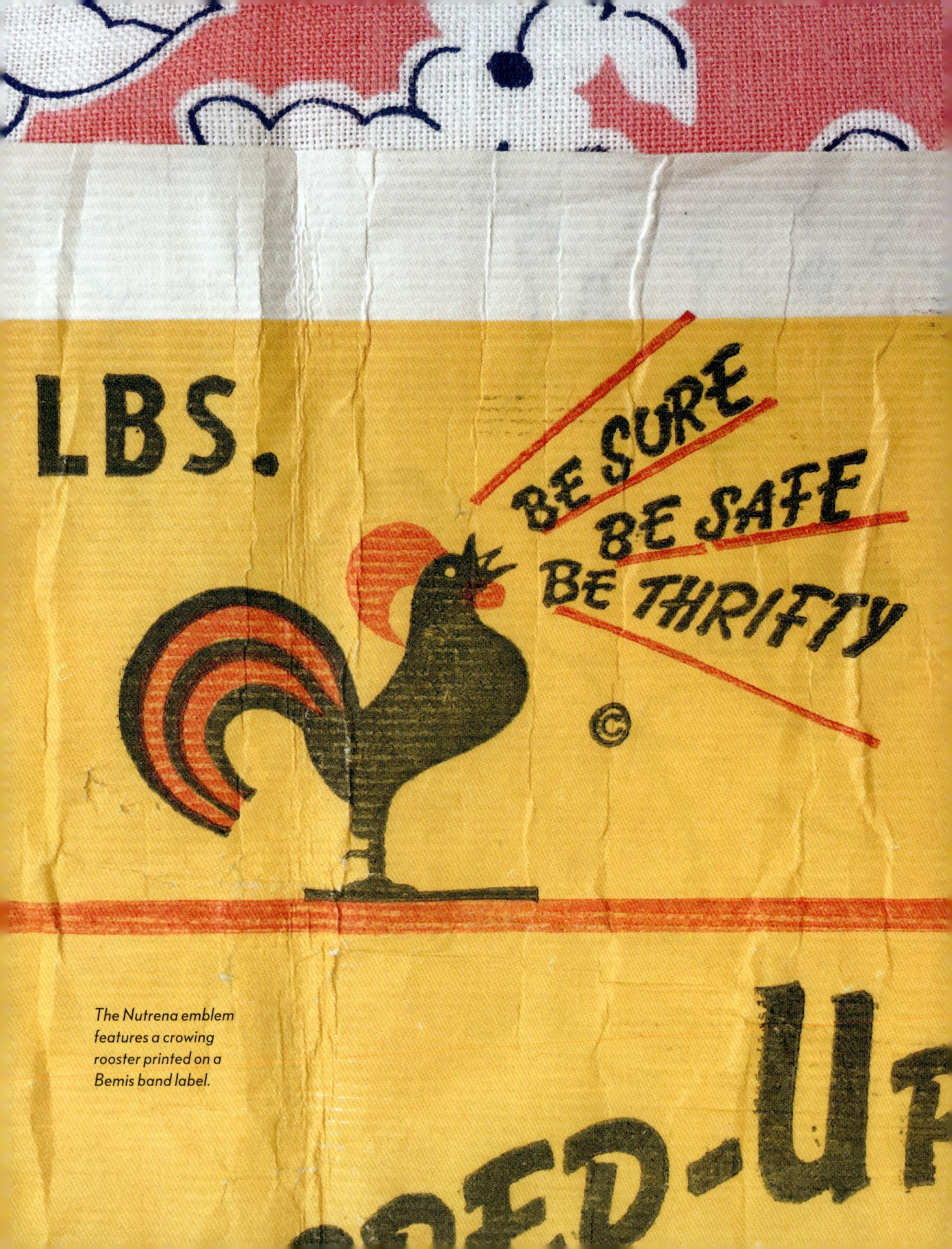
LBS.
BE SURE
BE SAFE
BE THRIFTY
©

The Nutrena emblem features a crowing rooster printed on a Bemis band label.

FOWL FASHION

It was not just feed and flour salesmen who took notice of the popularity of feed sacks. In 1949, the Poultry and Egg National Board (PENB) announced plans to "concentrate nation-wide attention on poultry products." How? By holding a contest to identify "the 10 best-dressed fowl in the nation." And these chicly clad turkeys, ducks and chickens could not wear just any old thing—their clothing had to be sewn from cotton bags that had contained poultry food (AKA "chicken linen"). In a September 3, 1949, article in the trade publication *Feedstuffs*, PENB general manager Homer Huntington drummed up enthusiasm for the promotion by declaring, "In the Fowl Fashion Show we have a contest that should arouse the interest of every news and photo editor in the country."

Although the concept might have seemed ridiculous, the first prize was anything but—a brand-new Kaiser automobile. The field was narrowed through regional and state competitions, and the top 10 fowl were brought for final judging to Kansas City. The second place winner, Priscilla Pinafore, owned by Mrs. Mary Pitlanish of Washington, Michigan, wore a polka dot and eyelet ensemble. An article in the January 26, 1950, *Charleston Gazette* from West Virginia noted that "Two of the three turkeys entered tried to put on an aerial show as a sideline" and that "'Daisy Duck' made the mistake of wearing a popcorn necklace. 'Model Future Biddy,' a hen, decided it was time for lunch and Daisy was left with a string around her neck."

MRS. MARY PITLANISH, OF WASHINGTON, MICH.
Costume for her hen wins regional contest

The grand prize went to Suzie Q, a hen attired in an 1830s-era French ensemble. The *Logansport Pharos Tribune* from Indiana noted that the "winning designer is Mrs. Micheline Riffle, Shenandoah, Iowa, a French war bride who has been in this country for four years. Mrs. Riffle combined French chic and American thriftiness in designing this cotton bag creation." Suzie Q donned her outfit again on February 16, 1950, during a three-day meeting of the Institute of American Poultry Industries in Kansas City, when the *Tucson Daily Citizen* from Arizona reported that the car was formally awarded to Mrs. Riffle, and once again for an ad in the May 1951 issue of *Feeds Illustrated*. "Slick Chick Wins First Prize in National Fowl Fashion Show," read the headline, above a photo that shows Suzie Q looking none too comfortable in her red dress and feathered bonnet.

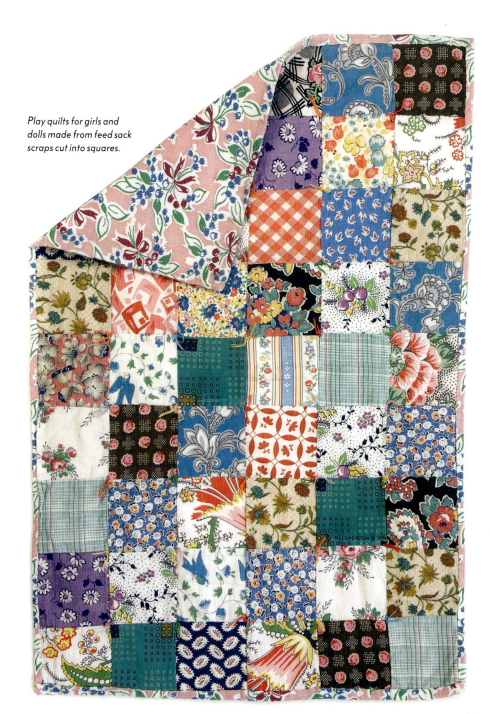

Play quilts for girls and dolls made from feed sack scraps cut into squares.

Pattern No. 8758—For the little "chicks" in the family, who think it fun to have play clothes just alike. Here is a "brother and sister" set to be made of sturdy cotton bags, fabric ready to stand the hard wear of playtime. Sizes 1 to 6 years. For sister, "chickie" dress, size 4 requires 3 bags 36 x42. For brother "chickie" suit, size 4 requires 2 bags 36x42.

CHILDREN'S CLOTHING

Feed sacks were especially suited to making simple children's clothes, and depending on the child's size, often required just a sack or two. Manufacturers recognized the popularity of children's clothes, and sacks patterned with juvenile motifs were plentiful. (Today they are also some of the most collectible.)

The National Cotton Council encouraged sewing for children through pamphlets—the mid-1940s brochure *Bag Magic* contained Pattern no. 8758, "for the 'little chicks' in the family, who think it fun to have play clothes just alike. Here is a 'brother and sister set' to be made of sturdy cotton bags, fabric ready to stand the hard wear of playtime." The shirtwaist dress for the girl required three 100-pound flour sacks, each of which yielded yardage that was 36 by 42 inches, while the overalls for the boy needed just two sacks.

When Florence Stockman was a child growing up on an Iowa City acreage, her mom made her pinafores and sun dresses from feed sacks, which she sometimes embellished with smocking. On her family's farm near Clarksville, Iowa, Phyllis Rosenwinkel's work clothes and play clothes were made from plain sacks that were bleached to remove the logo, or tinted using Rit dye. The family's chicken feed sacks, which featured pretty prints, were saved for clothes worn to school and other public places. Loretta Smith Steele of West Virginia had numerous halter tops made out of feed sacks: "They were easy to make and didn't take much fabric," she says.

Sunday, April 25, 1954

Flour Sack Made For Baby's Use

MEMPHIS, Tenn. (INS)—Putting baby in the sack has come to mean more than slang for tucking Junior in bed.

He literally is put in a sack—a cotton flannel sack, newest twist in flour packaging.

The National Cotton Council in Memphis reported that for the first time housewives can buy 25 pounds of flour in a high quality cotton flannel container.

Each bag provides a piece of fabric 27 inches square, suitable for making children's pajamas, bed jackets, baby gowns, and other sleepwear.

The sacks are available in five print patterns, each in a combination of two different colors.

The prints include a gray, red, and tan Wild West print, a toy animal design in pastel hues, and three dainty floral designs.

"Putting baby in the sack has come to mean more than slang for tucking Junior in bed," states an article from the Arizona Republic, *April 25, 1954.*

This selection of children's clothing made from feed sacks has been collected by Gloria Hall over the years.

DRESSES AND SEPARATES FOR WOMEN AND GIRLS

In an era of big box stores and fast fashion, it is hard to imagine a time when a woman's wardrobe consisted of one or two "good" dresses, along with a few items of clothing for work or play. These sartorial limitations likely made girls value the dresses they or their mothers stitched from feed sacks all the more.

Loretta Smith Steele wore feed sack dresses during her childhood in West Virginia. Her family had a cow, chickens and pigs, and raised much of their own food in their two gardens. "We were pretty self-sustaining," says Loretta. Her mother sewed shirtwaist dresses that required three feed sacks, and when she found three alike at the feed store she would buy one, and the shopkeeper would hold the other two for her until they were needed and could be paid for. Loretta's "Sunday dresses" were stitched from feed sacks. "I was very proud of them," she says. "We starched and ironed those things and thought we looked good. We were all in the same boat. Our dads worked away from home as well as having a farm. We were all in the same financial situation, we all wore the same kind of clothes, and we gave them to people who didn't have them. That's the way I grew up." When she was 18 and had her first job off the farm, her mother stitched her a sheath dress from feed sacks that were white with blue polka dots. Loretta wore it with white leather, ankle-strap sandals.

Farm Folk Prefer to Buy Feed Sold in Cotton Bags

MEMPHIS—Farm folk prefer to buy feed in cotton bags. Surveys at four state fairs show this to be true.

Cotton bags, because of their re-use value, appeal to farm wives. Almost 2,000 women were interviewed at state fairs in Minnesota, Iowa, Nebraska, and California.

More than 60 per cent of these women prefer that their husbands buy the 50 lb dress print bag. They stood by this preference even when told that the cotton bag costs 15 per cent more than other type bags.

More than 90 per cent of the women consider the dress print bag worth 20 cents to them. They set the value of the bleached cotton bag at 17 cents, and the unbleached bag at 15 cents.

The Britt News Tribune, Wednesday, Dec 5, 1951.

Girls in matching feed sack dresses at the Vermont state fair, Rutland. Photo by Jack Delano for the Farm Security Administration, September 1941.

LIBRARY OF CONGRESS

Nancy Freeland's mother stitched skirts for her and her sisters in the 1950s. "They were on a skinny band with a button—no zipper—that you could wiggle up on you," she says. Nancy has fond memories of those skirts. "Feed sacks were soft cotton, much better than those wool skirts we used to wear that would chew up our skin." She never felt there was anything socially unacceptable about feed sack garments. "It was a rural community and everyone did the same thing," she says. "There was no distinction between who had money and who didn't."

Not everyone has positive memories of feed sack clothing. Sandy Webrand's mother did not sew, and Sandy, who grew up in Iowa, wore ill-fitting hand-me-downs, or cobbled together outfits for herself from feed sacks. "My first feed sack dress was a dreadful sundress with a jacket—the pattern was self-drafted," she says. "The straps and bolero jacket were made of a purple feed sack print, floral, of course, and it was constructed on a hand-crank child's sewing machine. I was probably around 11 when I made it. I wore that dress to the Butler County Fair and got terribly sunburned."

Cotton Bags Are Key to Thrifty Sewing

Almost all of these bags come with labels. They come easily when soaked in water. If printed on, soak in warm, soapy water.

Use Cotton Bagging

Tea towels, dish towels, hand towels can be made in numerous colorful designs from the material that comes into your home in the form of cotton bags for flour, sugar, feed. Many designs can be embroidered with simple, easy-to-make stitches on the soft, white cloth that laundered cotton bags supply. Cotton bags make ideal dish towels because they are absorbent, durable, and leave no lint on the dishes.

Thursday, November 11, 1948.

IT'S IN THE BAG!

Gay red plaid cotton bags bring a note of sunshine to this suburban kitchen. Cotton bags were used to make the cheery curtains and matching tablecloth.

HOME FURNISHINGS AND FIX UP SECTION, WEDNESDAY, OCTO

Grandmother Wasn't So Dumb—Decorated Flour Sacks Will Make Pretty Curtains

- RINGS, PLASTIC OR BRASS, SEWED ON
- OTTAVIA CLIPS, EASILY SLIPPED ON AND OFF FOR WASHING
- LOOPS MADE OF MATCHING OR CONTRASTING COTTON BIAS TAPE
- BY-THE-YARD WASHABLE COTTON EDGING WITH LOOPS

COTTON bag cafe curtains may be hung in these carious ways.

FLOWERED bags make facing and valance on white curtains.

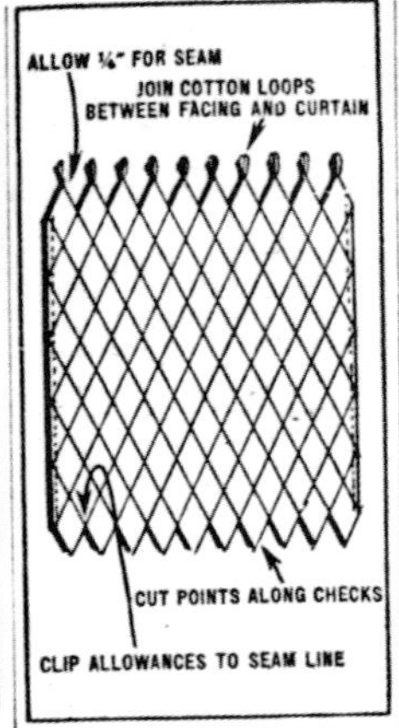

MATERIALS necessary:
One 100-pound bag for each panel of curtain, one for each valance, ½ yard of fabric for the contrast facing shown in the picture.
Two curtain rods for each window.
Rings, clips or specially made edging, if something other than matching fabric loops are desired (see sketch).
Thread, pencil, saucer, scissors.

HOW TO make them:
1. Install two rods in window, at top and center. For lower curtain, measure to sill, plus ¾ inch for finishing. For upper curtain (if this style, rather than the valance style, shown, is desired) measure from upper to lower rod.
2. Cut curtains to length determined. For valance, cut strip 13½ inches deep and another strip 4½ inches deep for the facing. For the contrasting facing shown here, cut two strips 4½ inches deep.
3. Fold curtain valance fabric

ALLOW ¼" FOR SEAM
JOIN COTTON LOOPS BETWEEN FACING AND CURTAIN
CUT POINTS ALONG CHECKS
CLIP ALLOWANCES TO SEAM LINE

DIAGONALLY checked bags make pointed cafe curtains.

lops, using saucer as guide for drawing them. Cut identical scallops in the top contrasting facing of the curtain and the facing for the valance.
4. Hem curtains at side edges. At lower straight edge and sides of the scalloped facing, turn material under ¼ inch and baste. Putting right side of this piece against the wrong side of the curtain, join the two scalloped edges, taking ½ inch seam. Trim seam to ¼ inch along curves and clip. Trim away entire seam at point. Turn facing to right side, press. Stitch edges and bottom of facing to curtain.
5. Baste one edge and both sides of bottom facing under ¼ inch. With right side of curtain, join lower edges, taking ¼-inch seam. Turn facing to right side, press and stitch sides and top of facing to curtain.
6. To join facing to valance, repeat step 4, but put right side of facing to right side of valance.

When Grandma bought her flour in 100-pound bags, she saw no reason to waste the empty bag. If she was lazy, she just cut three holes and Little Abagail went about dressed in it, labeled XXXX across the tummy. If she was energetic, she used the fabric for what she called sash curtains.

Today, sash curtains again come from cotton bags, but with two important differences—the bags are gaily printed in the most up-to-date patterns, and we call the curtains cafe curtains.

The National Cotton Council has prepared directions for making cafe curtains from cotton bags.

If 100-pound cotton bags aren't around the house in great numbers, the Cotton Council notes that they can be obtained through the mail order department of Sears, Roebuck and Company, bag firms and sometimes city bakeries.

Almost all of these bags come with labels. They come off easily when soaked in water. If printed on, soak in warm, soapy wa-

---

Sunday, November 1, 1953

Sacking Is Promoted As Decorative Fabric

By WALTER BREEDE JR.

NEW YORK (AP) — Lady, don't throw that flour sack away—it might make a pretty house frock or kitchen curtain.

That's the message the cotton industry is trying to put across in a stepped-up campaign to spur demand for cotton bagging.

The idea is to encourage manufacturers of flour to package their product in gaily printed cotton bags with a reuse value for home sewing. The same goes for producers of feed, fertilizer, and other products which were once put up in cotton sacks but now find their way to market sacked in burlap or paper.

PLANS FOR THE campaign were outlined this week at a meeting of the Cotton Bag Market Committee, representing textile mills, bag manufacturers, and cotton growers.

"Cotton bags are beginning to stage a comeback," said Herbert L. Kory, the committee's head. "Production showed a 10 per cent rise in September over August, and this may be the start of an upturn."

With paper cheap and burlap prices at their lowest level since early 1952 when India slashed its export tax on jute, demand for cotton bagging has slumped badly. This year's cotton bag consumption is estimated at 435 million yards. Two years ago the annual yardage was 480 million.

That's where the home-sewing gimmick comes in. The cotton people figure that if enough women want their feed or flour packaged in cotton bags that can be sewn into attractive household items, processors and distributors of these commodities will see the light.

"IT'S THE ONLY way we can compete with paper and burlap," one cotton spokesman said. "We can never compete on the basis of price alone."

A three-ply paper bag with a capacity of 50 pounds costs about 7 cents. This compares with 9 cents a yard for 7½-ounce burlap and 17 cents a yard for class-B cotton sheeting. It takes about 1½ yards of burlap or cotton to make a 100-pound bag.

In co-operation with sewing machine manufacturers, the cotton industry sponsored "save with cotton bags" sewing contests this fall at 33 state fairs. Some 10,000 women competed.

The final figures aren't in yet, but it's expected that home seamstresses will turn out 30 million dresses, pillowcases, and other items this year from an estimated 100

FLOWERCRAFT

How About Your XMAS?

Come in and let us help you with CORSAGES and ARRANGEMENTS of WOOD FIBRE FLOWERS

FREE CLASSES

ANN'S FLOWERCRAFT
1718 East McDowell Road
Phone AL 8-2225

STITCHING ALL THE REST (EVEN THE UNMENTIONABLES)

In many homes, nearly every item made of fabric, both utilitarian and decorative, was stitched from cotton bags. Feed sacks lived on as clothespin and laundry bags, hooked and braided rugs, shoe cases, dresser scarves, napkins and placemats, potholders, purses, bonnets, handkerchiefs and diapers. Ruth Rhoades learned that Irene Mixon made a cap from a feed sack in the 1920s to keep dust off her hair as she did housework. Mixon could not afford to purchase elastic to thread through the cap's casing, so she used quarter-inch strips of rubber cut from an inner tube to keep it in place.

Loretta Smith Steele says that during her West Virginia childhood, the pillowcases, tablecloths and sheets were adorned with lace and embroidery, "even though they were made from feed sacks." She still has a dresser scarf her mom made from a feed sack. "She crocheted the prettiest trim on both ends," says Loretta. "She didn't waste that fabric." Nancy Freeland's mother had a loom in the basement and wove rag rugs that likely included feed sack fabrics. Phyllis Rosenwinkel's mother stitched boxer shorts for her husband from feed sacks that she had soaked to remove the logo.

Debi Taylor of Harvest, Alabama, shared one of the lesser-known uses for feed sacks. Her husband was one of 15 children, and cotton bags played a prominent role in their home. Their father was employed as a farm labourer and none of his sons attended school past ninth grade, instead going to work on the farm. Debi's husband left school in seventh grade but started working much earlier, driving a wagon to the cotton gin when he was only six years old and so small that he had to stand up to reach the brakes. (He later returned to school, earning his GED diploma when he was 57.) Although his sisters were able to finish high school, they, too, lived frugally. Two of them, born in the 1930s, recall going to the feed store in a buggy, and as the bags were unloaded from wagons they would select matching feed sacks. Their mother made curtains and quilts out of the scraps left from the clothing she sewed, and the girls used strips of feed sack fabric during their menstrual cycles, washing them for reuse. Though disposable sanitary pads were available in the 1920s, their cost was out of reach for many women, who instead made use of what was available. The phrase "on the rag" derives from just such circumstances.

A cotton bag newspaper tip from 1953, above, and accessorizing with Little Things from the Sew Easy with Cotton Bags booklet, shown below.

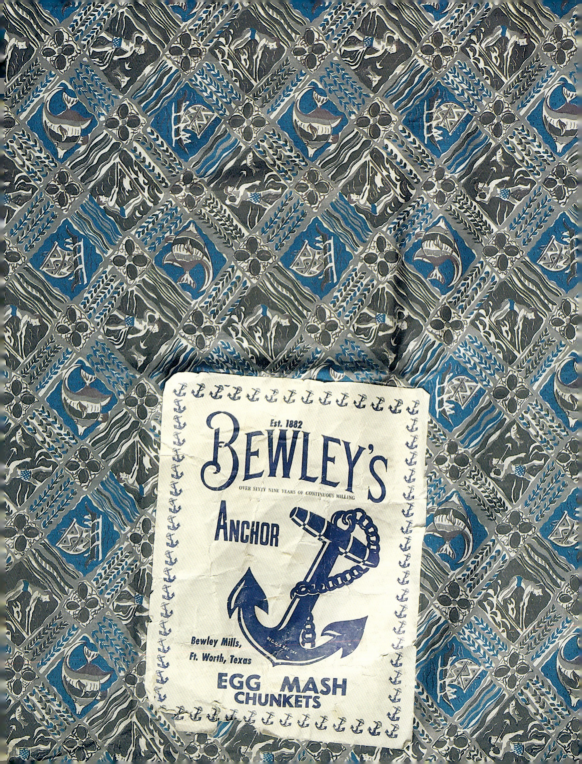

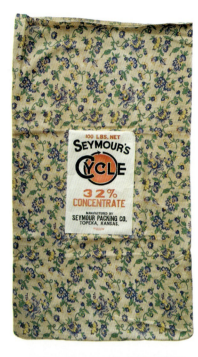

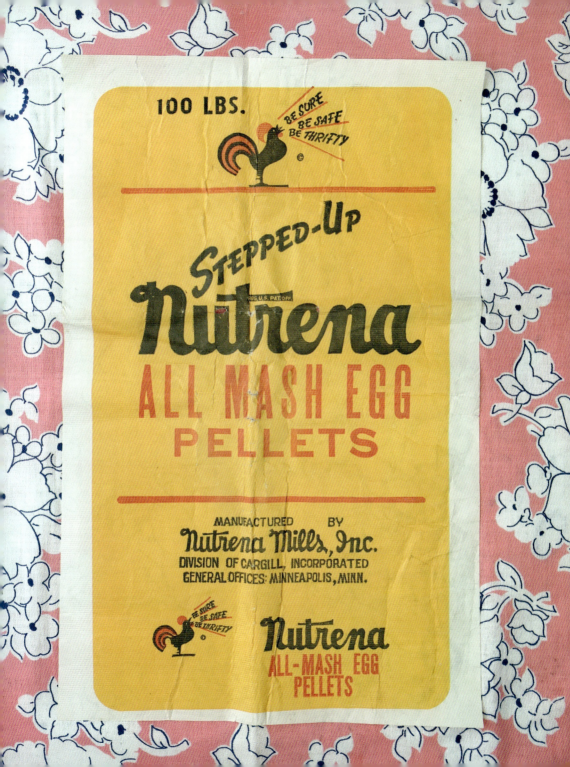

Coupons and special offers were sometimes sewn into the bottom seam of the bags. These coupons highlight the sewing potential of the sack material.

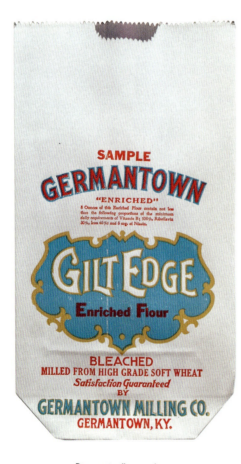

Dress print flour sacks competed against paper bags which were less expensive to produce and common in cities.

QUILTS

With carefully conserved feed sack remnants, women created quilts that rank as works of art. It has been suggested that the popularity of quilt patterns with small pieces, like Grandmother's Flower Garden or Double Wedding Ring, owed as much to quilter's propensity to use every little scrap as it did to prevailing fashion. (Carolyn Ducey, curator at the International Quilt Study Center and Museum in Lincoln, Nebraska, offers another possibility for quilters' love of patterns with tiny pieces: competition. Ducey notes that time and again in newspaper accounts of quilts, the total number of pieces was prominently cited. She speculates that quilters may have shared or traded pieces in order to create Postage Stamp quilts and other scrap quilts whose pieces numbered in the thousands.)

A crib quilt in an Axe Blade pattern by quiltmaker Joan White, circa 1940.

INTERNATIONAL QUILT STUDY CENTER, UNIVERSITY OF NEBRASKA-LINCOLN, 2011.003.0002

Nancy Freeland and her three sisters drove tractors, rounded up cows and scooped manure on her family's Ohio dairy farm. She remembers her mother making a quilt with four-inch squares left over from the skirts she made her daughters. "Nothing went to waste," says Nancy. Loretta Smith Steele would often find scraps from her Sunday dresses incorporated into her mother's quilts. When Loretta's younger brother was about 10 years old, he had a red and white feed sack shirt with roosters on it. Scraps from "Uncle Dave's rooster shirt" showed up over the years in quilts made by both Loretta's mother and Loretta, who inherited her mother's box of feed sack scraps. "She didn't throw away any of those tiny little pieces," Loretta says.

It is common today to discover feed sack quilts with tops that combine fabrics from a range of decades. Because quilting was by nature a way to conserve, scrap bins usually contained fabrics saved over multiple decades that were pulled out whenever a particular colour or pattern was needed. While some quilts were made entirely from feed sacks, many include feed sack material combined with other fabrics, produced both earlier and later than the feed sacks themselves: mourning prints, dress prints, shirting and others.

Colourful, intricately pieced quilts were not the only ones created from feed sacks. Utility quilts, made from plain sacks as well as sacks with visible logos, could be stitched quickly and were useful for those times when keeping warm overrode aesthetic interests: they might, for example, grace the bed of a hired hand who worked on a family's farm.

Feed sack collector Gloria Hall appreciates the ingenuity of this utility quilt in her collection— the top, made of four feed sacks, is an enlarged version of a nine-patch quilt block. The quilt measures approximately 76 inches by 76 inches.

In addition to being quickly stitched together, utility quilt tops were often tacked to their backings with big stitches of feed sack thread, embroidery floss or yarn, rather than being quilted.

Sometimes, as in the case of Sandy Webrand, who describes her farm upbringing in Butler County, Iowa, as "post-Depression and dirt poor," no one in the immediate family sewed. "My only blanket was a 'quilt' my paternal grandmother made for me," she says. "It was four feed sacks sewn together, dyed an uncertain shade of maroon and filled with a brown wool of some type. The layers were held together with two strands of yarn which formed an X." Her grandmother had not used the pretty, printed feed sacks, but rather ones made of coarse cotton, with company logos that remained visible through the dye. "It was waste not, want not," says Sandy. "These are memories I don't particularly relish."

Gloria Hall discovered one utility quilt that was sewn quickly, but that still featured the fresh colours of printed feed sacks. The quilt top is a giant nine-patch block, which its creator cleverly stitched from a whole sack, two halved sacks and one quartered sack.

Just as work clothes and undergarments were often stitched from plain feed sacks while printed sacks were saved for clothes worn to school and church, many feed sack quilts were also created with a similar mindset. White or off-white backing, created by stitching plain feed sacks together, was frequently used for the flip side of quilt tops pieced

A drawstring sack for Country Gentleman pipe and cigarette tobacco.

The "puff quilt" at right is stitched of tobacco sacks that were individually dyed, stuffed and hand-sewn together by Effie Roe of Kerr County, Texas, circa 1930s. The quilt is 56 inches by 81 inches.

WINEDALE QUILT COLLECTION, THE DOLPH BRISCOE CENTER FOR AMERICAN HISTORY, THE UNIVERSITY OF TEXAS AT AUSTIN.

out of patterned feed sacks. This "party in the front, business in the back" aesthetic represents an effort to create quilts that were spirited and cheerful, even when resources were limited.

Some quilters removed the graphics from the backing fabric, but for utility quilts, women often skipped this tedious step, with the assumption that the lettering and logos would fade after repeated washings. Other women could not resist elevating the humble materials they had to work with. When Effie Roe of Kerrville, Texas, created a tobacco sack quilt in the 1930s, she dyed and stuffed each of the more than 550 muslin tobacco sacks that she used before whipstitching them together. The resulting delicate, watercolour effect belies the fabric's rough origins. In her book *Kentucky Quilts and Their Makers*, Mary Washington Clarke writes that it was common to ask men to save these unbleached muslin tobacco pouches. She also mentions a Miss Evelyn Fuqua, who reported using 900 sacks in one quilt.

Plain sacks also served as foundation fabric for string-pieced quilts and crazy quilts. Dot Bloom's mother, who pieced her quilt tops by hand, used a sugar sack as the base of a crazy quilt she subsequently embellished with embroidery. Piecing on a foundation gives quilters a firm substructure on which to sew even the tiniest of scraps. And in the constant struggle to conserve materials and save money, older, ragged quilts were sometimes sewn inside newer quilts, in place of batting. (This practice was not limited to feed sack quilts.) The stuffing in utility quilts could include worn upholstery fabric or denim, or sometimes nothing at all. In *West Virginia Quilts and Quiltmakers: Echoes from the Hills*, author Fawn Valentine notes that some families would fill their quilts with burlap sacks: "Four burlap sacks … were sewn together and washed to make them soft, and used for filling. Sometimes two or three layers of burlap bags were used."

Long after feed sack production ceased, frugal sewists still used the sacks and scraps they had squirreled away to create quilts. When Loretta Smith Steele's mother passed away at the age of 95, Loretta inherited her box of feed sack scraps. "She didn't throw away any of those tiny little pieces, and I've used every single one in quilts," says Loretta, who says she enjoys sewing with them because they are soft and easy to piece and quilt by hand—Loretta's preferred mode of stitching. But the bottom line for her is that using feed sacks reminds her of her childhood: "I wash them and hang them on the clothesline to dry and they smell so good—like fresh air. They smell like summertime and my childhood."

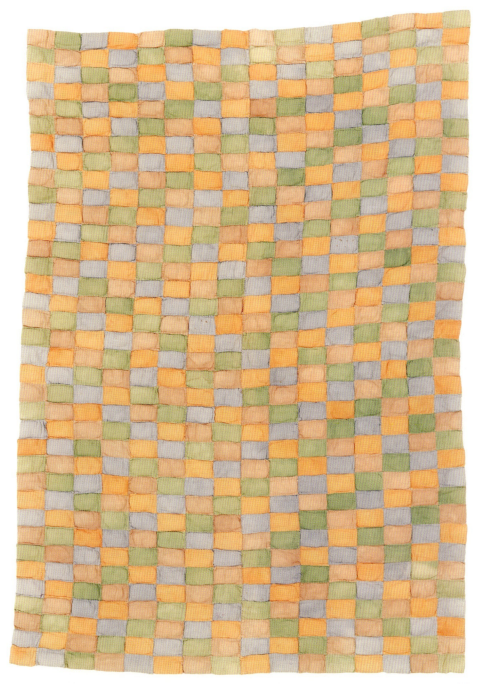

Scrap Bag Ideas...

Your scrap bag is chock-full of ideas waiting to be brought to life. Use needle magic to turn odds and ends into clothing accessories, household items, and gifts for family and friends. To help you get started, we offer the suggestions on these pages.

Piece A Quilt

Quilts symbolize the thrift, patience, and ingenuity of the American women of earlier generations. They can be works of art, treasured by the family and handed down from one generation to another.

Old quilt patterns, some of them masterpieces of needlework, are fashion-wise today. Here are two traditional designs, one simple and the other a bit more difficult. Both are well worth your time and effort.

Salvage your most vivid bag prints for quilt pattern called Joseph's Coat, *upper right*. Use diagram for cutting pattern parts.

Full-blown Tulip, an interesting pattern design in five pieces as diagrammed *lower right*, is pretty enough to use as a bedspread.

Dust ruffles, made from cotton bags, attractively complement quilts.

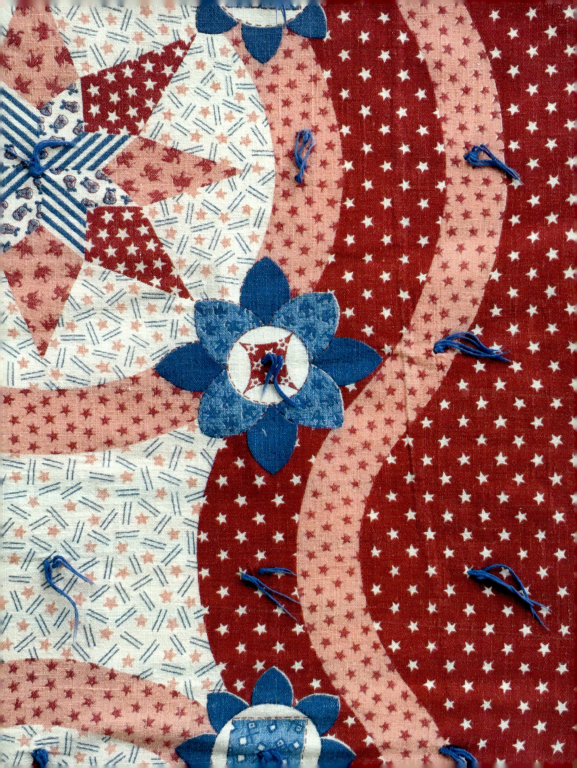

This crazy quilt's top was made with delicate silks and dress fabrics foundation-pieced onto flour sacks.

FEED SACKS AROUND THE WORLD

Reusing cotton sack fabric undoubtedly occurred everywhere that cloth bags were found—and indeed, sack fabric has shown up in quilts from Canada to Ireland to Australia. These were largely cotton bags used for flour—plain fabric with printed logos. Whether due to poverty and desperation, or simply a sense of economy, throwing away "free fabric" was unthinkable.

Dress print fabric bags were most likely produced only by United States-based manufacturers, although some were sold in Canada and others undoubtedly travelled farther afield via friends and family. Today, vintage feed sacks are bought and sold internationally and they can be found wherever there are quilters: Japanese quilters are so fond of feed sacks that the bags can bring up to several hundred dollars apiece there.

European grain sacks, handwoven in the 1800s, are also collectible and valued by decorators and sewists for their subtle textures and tones. Like the early cotton bags used in the United States, European grain sacks were handwoven and handsewn, and used by farmers to take grain to the mill and to bring home flour, cornmeal and other foodstuffs. Kaari Meng, proprietor of French General in Los Angeles, has collected and studied European textiles for 25 years. She says that French sacks were typically embroidered with the owner's initials, and sometimes a number, to ensure that the valuable bag and its contents would return to the proper owner. German sacks were more likely to be stenciled with a tar-based ink, and to include more information—the owner's first and last name and perhaps his title, the name of the town nearest the farm and sometimes the year. "They provide a good peek into European history and have stood the test of time," Kaari says. "The heavy fibres they used make the bags almost indestructible, and they were probably passed from generation to generation."

She believes that in some regions grain sacks were included as part of a bridal trousseau, and she has seen examples of alterations to the initials embroidered on the sack, reflecting the bride's new married name. In addition to embroidering and stenciling ownership information, stripes were often woven into the fabric of the sacks, and their colour and placement was likely unique to a particular farm or owner. As in the United States and Canada, sacks had other uses: Kaari finds grain sacks with military and postal markings at European flea markets.

THE END OF AN ERA

A 1960 survey of rural women showed that more than 50 percent still sewed with cotton bags, and the McCall Pattern Company produced patterns especially for bag sewing as late as 1961. Despite this interest and the creative marketing efforts of cotton bag manufacturers, cotton feed sack production wound down around 1964. Not only could cotton bags not compete with the lower cost of paper, purchased clothing was becoming within reach of more families during this period of prosperity, and home sewing declined as opportunities increased for women outside the home.

The exact number of prints produced in the cotton bag era is impossible to determine. Collectors Pat and Cecelia Reid have documented 18,000 different prints and prints in multiple colourways, but Cecelia admits, "We quit counting at 18,001." Collectors Gloria Hall and Paul Pugsley estimate the number of different prints is in the tens of thousands, though Gloria acknowledges that the exact number may never be known. A conversation that she had with a longtime salesman for Percy Kent reinforced her suspicion that the exact records of many patterns are simply unattainable. "Percy Kent produced their own prints and he told me that when the company was sold, all those records were destroyed," she says.

Though a few items are still sold in cloth bags, these are largely produced in small quantities for limited markets. Most contemporary feed and seed bags are made of polypropylene plastic rather than cotton, but resourceful folks still repurpose them. The Internet abounds with instructions for using these water-resistant bags to make everything from the tried-and-true apron to wind breaks, tarps and container gardens. Asian rice sacks are often used to create messenger bags.

Exit Flour Sack, Once So Useful

BOSTON (UPI) — The flour sack which poor folks used to make into bedsheets, dish towels and even clothing, is on the way to oblivion, according to Seaboard Allied Milling Co.

Says President Otto Bresky, "Flour for large consumers is destined to be handled more and more in bulk in special trucks and in paper containers for the housewife."

The Times Recorder of Ohio briefly commented on the demise of the flour sack on August 21, 1960.

FEED SACKS TODAY

Today, more than 50 years after the last printed sacks rolled off the production line, quilters and other textile lovers still appreciate them, and companies continue to create reproduction lines of feed sack fabric, as well as fabric lines inspired by those prints.

Cheryl Freydberg, vice president and design director at Moda Fabrics, loves feed sacks for their variety and graphic quality. "They appeal to a wide audience," she says. "The smaller, sweet, ditzy prints appeal to one audience and the more graphic, bold sacks appeal to a more contemporary, modern quilter." Cheryl finds that the simple, fast production methods used to create feed sacks adds to their graphic appeal: "With some reproduction fabrics, you have layer upon layer of detail, while these seem to have been printed more crudely and simply, with only about six colours in the average sack." She also appreciates the tremendous variety of prints in feed sacks. "There's nothing really complicated or fancy or fussy about them," she says.

Mickey Krueger, president of Windham Fabrics, has been collecting feed sack fabrics and quilts for the company's archives for more than 20 years, and the collection serves as the basis for at least three feed sack reproduction or design-affiliated lines a year. Mickey finds that both traditional and modern quilters are drawn to feed sack imagery, and while he appreciates the visual appeal, he is equally drawn to the story behind the bags. "I love their history and I love the diversity of ways people used them," he says. "I especially love what they mean for us in the quilting arts today. It is a great American fabric story."

That fabric story appeals to quilt and fabric designer Denyse Schmidt as well. "I love that quilting, in its deepest roots, carries a sense of reuse and of the preciousness of every last bit of fabric," she says. "It isn't our reality today, but it really resonates with me and speaks to not wanting to collect or have excess, but of using up what you've got."

At left, the Chinese Lantern colourway of Denyse Schmidt's fabric collection entitled Hadley, which showcases "smart, pretty florals, confident plaids and universally useful calicos." The full collection is shown above.

Inspired by this vintage dimensional star made from feed sacks, Amy Barickman of Indygo Junction has developed a line of dimensional paper piecing kits.

Denyse's introduction to feed sacks came while trolling through flea markets, and she was drawn first to the way the sacks were printed. "I assume the printing process was inexpensive—maybe the cotton shifted a little—and that quality of design is part of the aesthetic I love so much," she says. Denyse sometimes replicates the feel of a feed sack print's slightly wonky registration by adding a bit of white behind a dot in her own fabric designs. "Printed matter that has this same quality—carnival and circus posters, for example—appeals to me," she says. "It's not the highest-end printing, but it serves its purpose really well and that's something I appreciate."

Amy Barickman uses feed sacks as the basis for projects that she creates for her company Indygo Junction and for some of the fabrics she designs. "I could sit and play with feed sacks all day long, dreaming up ideas," she says. She, too, appreciates the story behind feed sacks. "In the past, sewing and quilting were about economic survival, and women's lives revolved around them—sewing, quilting and dressmaking were such a huge part of our female culture," she says. "You think about the hardships of the family farm, but also about going to the feed store and picking out a print you knew would become your child's sunsuit, and the joy and personal fulfillment of sewing it." Many of Amy's fabric lines contain at least a couple of reworked feed sack prints, changing the scale of the imagery and recolouring the designs. She finds many feed sack prints work well for contemporary clothing, as well as quilting. "Trends come and go, but the graphic quality of feed sacks and the incredible variety of patterns makes their appeal timeless," she says.

That timeless visual appeal, combined with the story feed sacks tell of the ingenuity of women (and marketers), is often viewed sentimentally, but it can also be instructive. In our era of abundance, "living simply" has become a mantra of contemporary life, and yet doing so often requires disposing of the things we have acquired.

How much better it would be if we were able to use up what we have, to make it work, to make it last, to reuse it so that nothing is wasted. Dot Bloom remembers the pride her family took during the Depression in creating a child's article of clothing by taking apart an older sister's outgrown skirt, turning it over and using the less worn side. There is satisfaction to be gained in using up every last bit.

Many people interviewed for this book acknowledged hard times, but also spoke of their enjoyment in learning from their mothers and fathers how to make a home with what they had, not with what they bought. The opening paragraph of the sewing pamphlet *Bag Magic* suggests (albeit in a rather grand fashion) that passing on just that kind of knowledge is an educational imperative: "It is [our mothers and teachers] who instill into the warp and woof of daily life the simple virtues of thrift, ingenuity and skill—the virtues upon which, in the last analysis, the future ... rests."

Reusing feed sacks was a way to survive, but it also allowed families to thrive. Making use of what they had was not based on assuming some high moral ground; it was simply a way of life. Feed sacks also proved to manufacturers that there was value in products whose reusability was enhanced. These are perspectives worth pondering today.

HOW TO TELL IF IT'S A FEED SACK

"Bags of all Kinds"

The easiest way to know if the vintage fabric you have is a feed sack is if it is still sewn into a bag shape. The selvedges are typically at the "mouth" of the bag so that it won't ravel—the raw edges are stitched together to form one side seam, while the other side is simply a fold. Large sacks (100-pound sacks, for example) may have selvedges at the top and bottom of the bag, indicating that the manufacturer used the entire width of the woven textile to make the bag.

When determining if a flat piece of fabric was once a feed sack, look for the stitching holes, where string previously held the bag together. Because feed sacks were stitched with heavy-duty thread or string, the holes will persist, even after the bag has been washed.

Serious feed sack collectors swear they can tell whether a piece of fabric was once a feed sack based on their familiarity with the weaves and patterns. Because the same fabric was sometimes sewn into bags and sold as on-the-bolt yardage, this can be challenging. According to collector Paul Pugsley, fabric for both feed sacks and yardage was generally 36-inches wide, although depending on the manufacturer and mill this could range from 35 inches to 38 inches.

Determining whether or not a piece of vintage clothing was stitched from a feed sack can be difficult, but stitching holes in the fabric are one clue. Some sewists avoided using the sections with holes, but those who maximized their available fabric would sometimes hide the holes in seam allowances, so inspecting them closely may reveal a garment's origins.

Is that quilt actually pieced from feed sacks? Again, collectors who are familiar with numerous feed sacks may be able to ascertain the origins of particular fabrics in a quilt. Because so many feed sack quilts are scrap quilts, it is common to find true feed sack fabrics mixed with both older and more modern fabrics: quilters kept every little bit and even passed their scraps along. Feed sack scraps could feasibly be combined in a quilt with scraps of turn-of-the-century mourning prints and 1970s dress prints. There are also people making quilts with feed sacks today, so a quilt stitched from vintage feed sacks could have been made last year. It is possible, according to collector Gloria Hall, for a quilt appraiser to examine thread, batting and stitching to distinguish a modern-made feed sack quilt from one constructed in the 1930s or 1940s.

Plain feed sacks were used for quilt backs, borders, sashing and even binding. Check closely for "ghost" images—lettering or logos that persisted despite repeated washings. These can also be found on some clothing, particularly undergarments or children's play clothing.

The bottom line, according to Gloria Hall, is that "a lot of knowing whether it's a feed sack is just being familiar with the prints, and also verifying with another expert. There's not one way to tell."

Use the tables below for checking bag sizes:

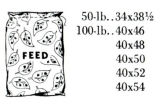

FEED
50-lb...34x38½
100-lb...40x46
40x48
40x50
40x52
40x54

SUGAR
5-lb...13x16
10-lb...16x21
25-lb...22x27
100-lb...36x40

FERTILIZER
100-lb...36x39
125-lb...36x45
150-lb...40x44
200-lb...40x52

FLOUR
5-lb...15x19
10-lb...18x23
25-lb...26x26
50-lb...30x34
100-lb...36x42

SALT
5-lb...13x14
10-lb...16x17
25-lb...18x26
100-lb...30x36

MEAL
5-lb...15x16
10-lb...18x22
25-lb...26x27
100-lb...36x44

This C and H sugar sack includes a copyright date of 1934.

DATING FEED SACKS

Determining when a feed sack was made is difficult and often impossible. Mills that produced fabric for bags have been closed for years and records have been lost. Here are a few clues to help you date a feed sack.

The amount of feed or other product that a sack contained, typically stamped on the sack or on a paper band, can provide a clue to its date. The sizes of early feed sacks were based on barrel weights: a standard barrel contained 196 pounds and bags were sold as quarter-barrel bags (49 pounds) and half-barrel bags (98 pounds). Smaller bags in weights of 12, 6, 4, 3 and 2 pounds were popular for home use. In 1943, the United States War Production Board standardized bag sizes based on the hundredweight and limited them to five sizes—2, 5, 10, 50 and 100-plus pounds, thereby streamlining production and encouraging manufacturers to focus on production for the war effort. Bags with odd-numbered sizes, therefore, were manufactured prior to 1943.

Some bags have a copyright date printed on them.

If a bag is stamped with a blue eagle and the letters NIRA, the symbol for the National Industrial Recovery Act, this is a clue that it was made between June 16, 1933, when the Act was signed into law, and two years later, when it was declared unconstitutional, according to author and collector Edie McGinnis. The goal of the act was to regulate industry and stimulate economic recovery.

Bags with logos from small mills (of which there were hundreds) may be dated by doing research into the mill where the product was produced—local history museums, newspaper archives and census records may be helpful. The *Northwestern Miller*, a trade newspaper published from 1873 to 1902, includes ads for hundreds of mills. Archived in-house publications for some of the major bag companies, such as Bemis's *Bemistory* or *Feedstuffs*, a feed industry publication, may offer information on some bag patterns. These publications are available through public and university libraries, historical museums and online. Feed sack ephemera—advertisements, brochures and newspaper articles—depict sacks and items sewn with them. Matching a sack to a print in an ad is an obvious way to date a sack.

Flour enrichment was initiated in the United States in 1938, which means that sacks printed with the word "enriched" can be dated from 1938 or later.

Sugar sacks that are printed with the words "A product of Cuba" are from pre-1960. Previous to that time, the largest portion of sugar imported to the United States came from Cuba, but politics changed that. Few sacks were produced after 1960, however, so this only helps identify a limited number.

The popularity of feed sack sewing led to bag fabrics also becoming stylish, so fabric designs and motifs can be presumed to be similar to designs that were popular in store-bought clothing of the same era. Comparing the colours, imagery and size of motifs may help in assigning an approximate date to a feed sack. Collector Gloria Hall says that the imagery on early feed sack fabrics, from the late 1920s and 1930s, was often outlined. Images of a very small number of feed sacks can also be found in newspaper stories and advertisements about feed sack sewing.

Bags with imagery that pertain to historic events (World War II, for example) can be dated approximately, based on those events. The same can be said for bags with imagery from television shows (such as *Davy Crockett*, which aired from 1954 to 1955) or movies (*Gone with the Wind* feed sacks were printed in three different colourways—the book was published in 1936; the movie opened in 1940). Edie McGinnis has a sack with Li'l Abner on it that likely corresponds to the newspaper cartoon strip's publication dates, beginning in 1934 to the early 1960s, when feed sack production ceased (though *Li'l Abner* continued to be published until 1977). Edie notes that certain particularly collectible prints, like the one with scenes from *Gone with the Wind*, were released in different colourways at six-month to one-year intervals, allowing each colourway to sell out before introducing another.

This project was foundation-pieced onto newspapers. Its mix of fabrics reflects a typical scrap bag and likely contains fabrics of multiple eras, including feed sacks.

Further investigation of the newspapers reveal articles about the 62nd Kentucky Derby, slated to run the following day. This dates the newspaper to May 1, 1936, and one can surmise that some of the fabrics are from a similar period.

WASHING FEED SACKS

Most dress print feed sacks will hold up to washing because the inks used to print them were, for the most part, colourfast. Nevertheless, it never hurts to test a corner of a sack, particularly one with bright colours, to make sure the colours won't bleed. In her book *Sugar Sack Quilts*, author Glenna Hailey recommends wetting a white cloth and dabbing it on the sack to see if the colour transfers. If it does, she suggests washing that sack by itself—otherwise, multiple sacks can be washed together.

There are nearly as many suggestions today about the best method for laundering feed sacks as there were recommendations for removing their logos in the 1930s. Hailey recommends soaking the fabric in an oxygen-based cleaner for 24 hours, then washing it with detergent. Collector and historian Gloria Hall endorses using a cleaning product that contains sodium perborate granules, which she says is safe

for most textiles and available at quilt shows. Hall soaks sacks in her bathtub, first tossing in a couple of handfuls of sodium perborate and agitating it before adding the fabric. She has soaked feed sacks up to six days, changing the water daily. Feed sack collector and quilter Loretta Smith Steele has the most cavalier attitude toward cleaning her sacks: she tosses them in the washing machine with detergent and bleach. "You cannot hurt a feed sack," says Loretta, who dries her sacks on the clothesline. "The first ones I bought I thought I'd have to treat with TLC, and I probably did for awhile, but they had stains and eventually I used bleach and they came out well."

Use caution if you decide to launder a bag with a logo on it, because many were printed with inks designed to disappear when soaked in water. And do not wash a feed sack with a paper label if you value it as a collectible—labels increase their market price.

Washing a quilt is another story. Because so many quilts sewn with feed sack fabrics also include fabrics from other eras, it is quite possible that some fabrics may bleed (red is notorious for this). The batting may shrink or bunch up—it is often difficult to tell what is between the top and backing of an older quilt. If a quilt is intended just for display purposes, using it "as is" may be the safest bet. If a quilt absolutely must be washed, give Gloria Hall's bathtub-and-sodium-perborate method a try after testing it with a wet, white cloth—but be aware that a formerly white-and-red quilt may become pink.

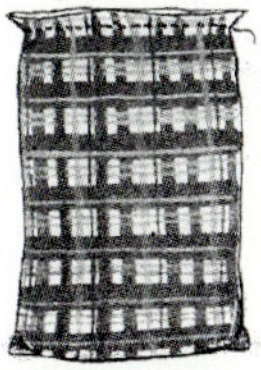

Here's How to Use COTTON BAGS...

This is the way to:

RIP THE BAG

Cotton bags are sewn with a chain stitch. Cut chain close to bag in corner. Take top thread in one hand, bottom thread in the other. Pull ... that's all there is to it!

REMOVE THE LABEL

Almost all bags have band labels for brand identification. Soak the bag in water and the label comes off in a jiffy! Some brand names are printed in washable inks that come out easily when soaked in warm, soapy water.

COLLECTING COMPANIONS

A love of history and "pure curiosity" are the forces that drive feed sack collector and researcher Gloria Hall. A longtime quilt appraiser, Gloria shares her knowledge and skills with guilds and at county fairs close to her home in Palmyra, Nebraska, and helps document fabrics and quilts for the International Quilt Study Center and Museum in nearby Lincoln. "What I like most are historical facts and historical anything. I'm very inquisitive," says Gloria, who gives talks internationally and around the US, and helped mentor students who today serve at quilt museums around the country.

Gloria grew up on Long Island in New York State and first learned of feed sacks at Fanning's Granary in Hampton Bays. There she and her sister would choose feed sacks for the dresses her mother and grandmother stitched. "Both were big seamstresses and made suits, coats, evening gowns and wedding gowns," says Gloria. "If we wanted something to wear, they got the fabric and they made it. For us, going to the granary was like kids today going to the mall." Gloria and her sister would select one feed sack, then change their minds and choose another. "We drove Mr. Fanning crazy, but he was a very quiet, complacent man who let us run all over."

Though the feed sack memories of her youth are fond ones, it was not until she married her husband in 1973 and moved to Nebraska that she caught the bug. Warren Hall was a rancher and farmer, and on weekends he and Gloria attended auctions, where Warren admired the feed sacks he remembered selling to customers at an implement store in Mound City, Missouri, where he had worked as a young man. His enthusiasm was contagious and Gloria started buying up feed sacks. She sold them for 18 years at major quilt events around the country and in Japan. One of her largest purchases was in 1999, after speaking at the Smucker's brothers' family reunion in Lancaster, Pennsylvania. A local man approached her about buying 2,500 sacks. "He was single and his folks had a hatchery and his mom had saved feed sacks through the years," says Gloria. She negotiated a price, then rented a U-Haul to cart the sacks home. "When I told my husband what I'd done, he said, 'It's like winning the lottery!'" Gloria bought them for $7 apiece and the first one she sold was a nursery rhyme-themed sack that brought $150.

Just as Warren shared his enthusiasm for feed sacks with Gloria, she has passed on her zeal to her grandson Paul Pugsley. Paul helped Gloria take feed sacks to shows and lectures after Warren passed away in 2007,

and has amassed a significant collection of his own. He has a special focus on feed sacks with transportation themes—cars, planes, ships and trains. In addition to having a love of feed sacks, he has become an expert in his own right, and the two collectors bounce feed sack facts and figures off one another with ease.

Although Gloria stitches quilts with her duplicate feed sacks, they have a far greater meaning to her than as just raw material. She loves that feed sacks represent the "explosion of colour" made possible after the United States, England and France regained access to dyes from Germany after World War I. She is fascinated by the quilts, aprons, towels and dresses made from feed sacks, each of which represents a woman's time and talents. "Each is different because each person is different, and we all see differently," says Gloria. "It's not that you'd hang them on a wall, but they are pieces of art."

You would think Gloria has seen it all during her more than 25 years of feed sack collecting, but she still gets excited about unusual finds. One of her current favourites is a quilt that has been stitched from four sacks. It is essentially a giant nine-patch block, with one 100-pound sack serving as the centre, bordered by two more halved sacks. The four corner blocks are another, quartered, sack. The back is made of feed sacks stitched together, and the front and back are stitched together like a giant pillowcase. The result is a utility quilt, but the soft hand and brightly coloured fabrics make it a pleasure to lie under on a cool summer evening. For Gloria, feed sacks never get old.

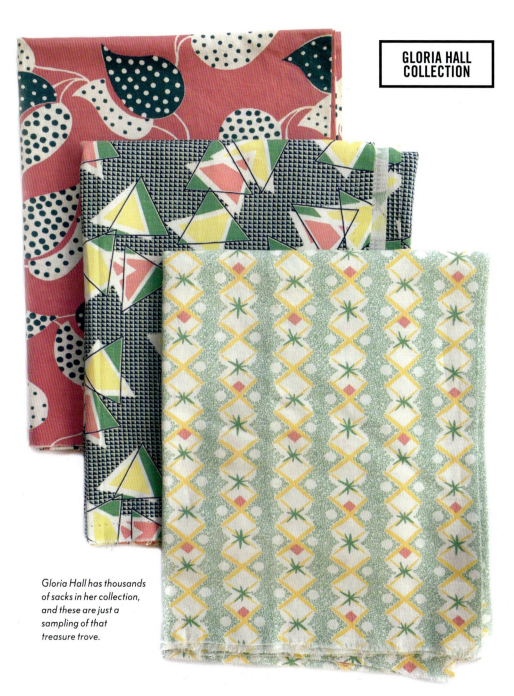

GLORIA HALL COLLECTION

Gloria Hall has thousands of sacks in her collection, and these are just a sampling of that treasure trove.

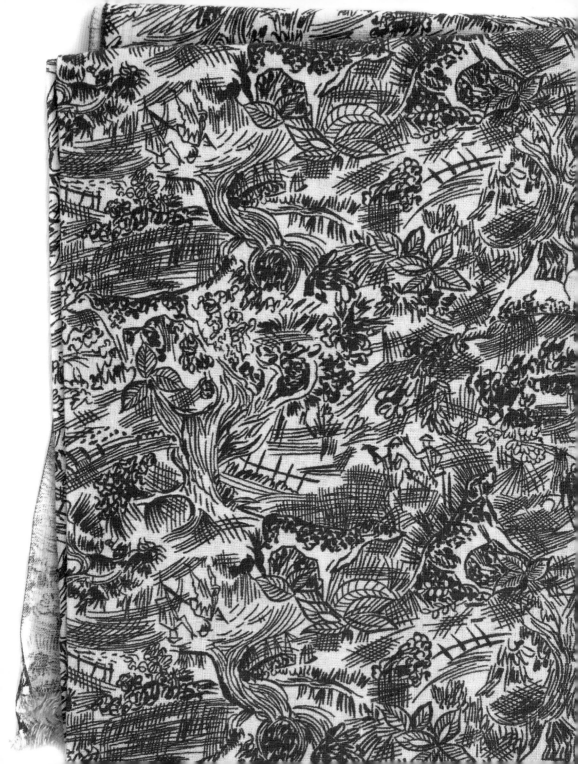

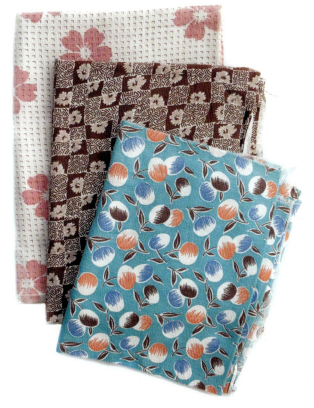

PAUL PUGSLEY COLLECTION

With a collection rivalling that of Gloria Hall—his grandmother—Paul Pugsley has an enviable selection of dress print bags.

From florals of all kinds, pictoral scenes, novelty prints and stripes—the beauty of feed sacks is enhanced through their juxtaposition.

Novelty prints include boyish adventure themes and transportation of all kinds.

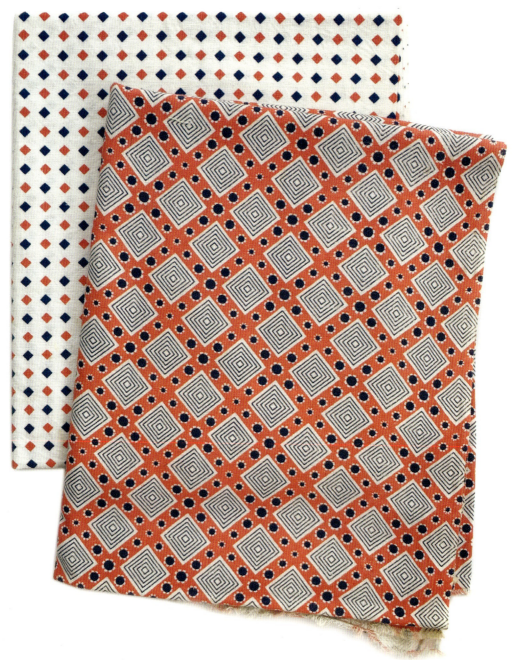

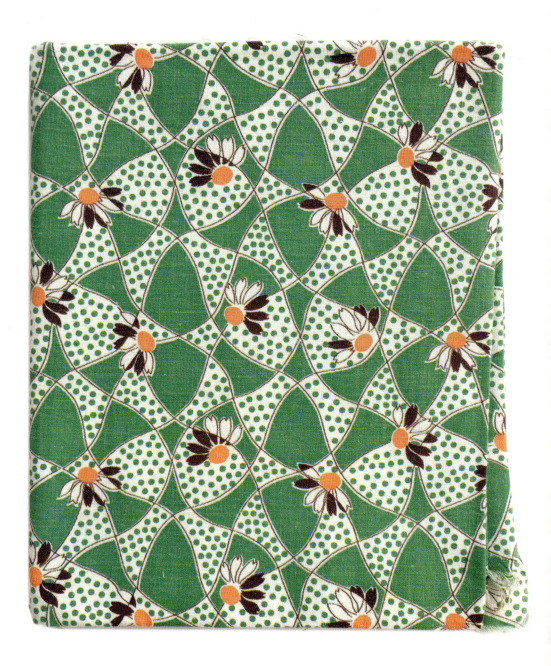

CHARLENE BREWER'S FEED SACK NOTEBOOKS

It started on a whim. Charlene Brewer bought some feed sacks, perhaps because they reminded her of the quilts her grandmother made. Then she started swapping swatches with feed sack collectors from North America and around the world. Soon she was making feed sack quilts, learning about the history of feed sacks, and giving programs about them to guilds and groups locally and across the country.

In the 1990s she became intrigued by the variety of prints and began filling notebooks with swatches, organized by colour and theme. She would snip a bit of fabric from each sack she collected and paste them in the notebook. When she bought a new sack she would double-check its pattern against her swatches to be sure she was not buying duplicates. Charlene took the notebooks to her lectures and to quilt shows, and people would stop and page through them all. "They'd be looking for something special, a dress or shirt fabric they might have had when they were young," she says. Charlene stopped adding to her notebooks around 2008. "When I started, I had no idea I would collect so many. There really is no end to it."

In 2010, Charlene donated a set of 12 three-ring binders filled with hundreds of swatches to the International Quilt Study Center and Museum (IQSCM) in Lincoln, Nebraska, whose mission is to "uncover the world through the artistic and cultural significance of quilts, and to research, acquire, and exhibit them in all their forms and expressions." IQSCM volunteers carefully removed the swatches from the notebooks and then stitched them onto archival paper. In addition to grouping similarly coloured swatches together, Charlene organized the 12 binders by categories that included "kitchen," "butterflies," "trademarks" and "novelty." They are available for researchers to examine, by appointment. Paging through this eclectic collection is a great reminder of the range of feed sack designs, from sweet, ditsy florals to Atomic Age motifs, and everything in between.

The ICSCM's feed sack quilt collection is just a tiny portion of the museum's holdings, which date from the 1700s and include quilts from 50 countries. The IQSCM was established in 1997, and in 2008 moved into a 37,000-square-foot building designed for them by Robert A. M. Stern. A 2015 addition doubled the museum's gallery and storage space. Work at the IQSCM includes quilt conservation, and researchers may use the museum's education collection and research materials by appointment. Visitors can view a variety of rotating exhibitions in the museum's galleries. The IQSCM has the world's largest publicly held quilt collection.

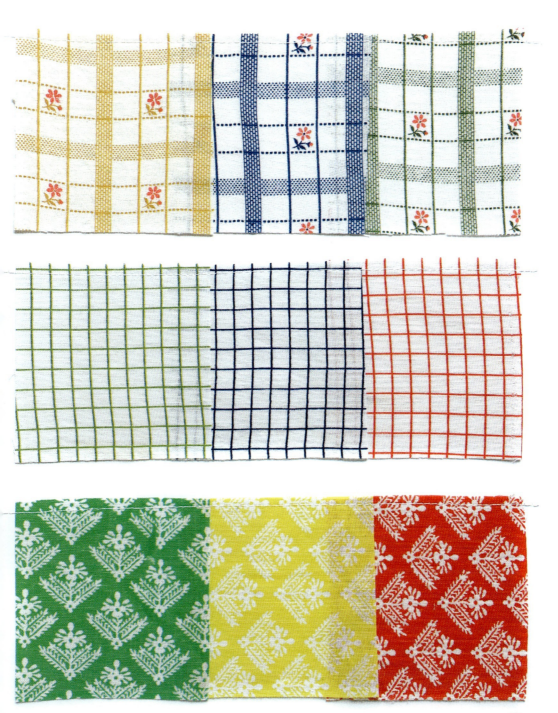

SCRAPS AND SWATCHES

The swatches on the following pages are from the collection of *Feed Sacks* book designer and UPPERCASE publisher Janine Vangool (who amassed hundreds of scraps over the course of creating this book) and collector Sharon Forth. Like many quilters, Sharon Forth has an ever-growing fabric stash. Of special interest are her hundreds of feed sack swatches and fat quarters, which she generously lent to us to photograph for this book. Although Sharon never wore clothing made from feed sack dress prints, she recalls her mother bleaching sugar sacks and decorating them with fancy work.

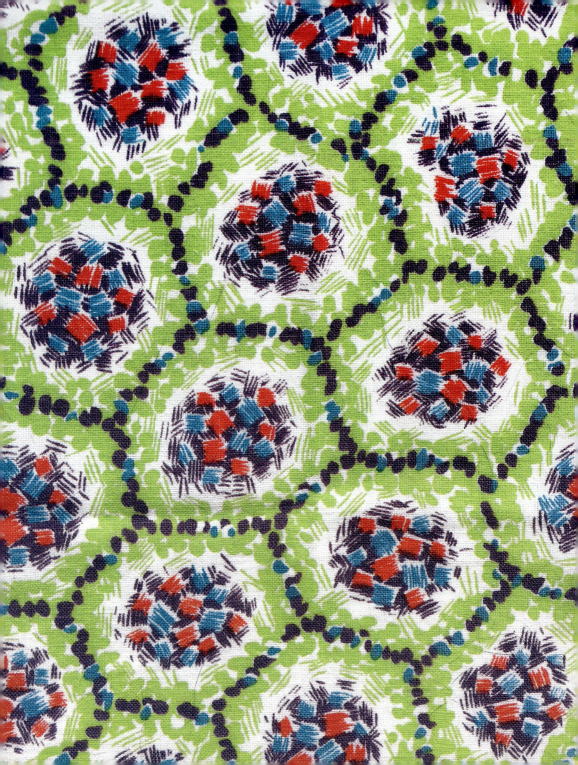

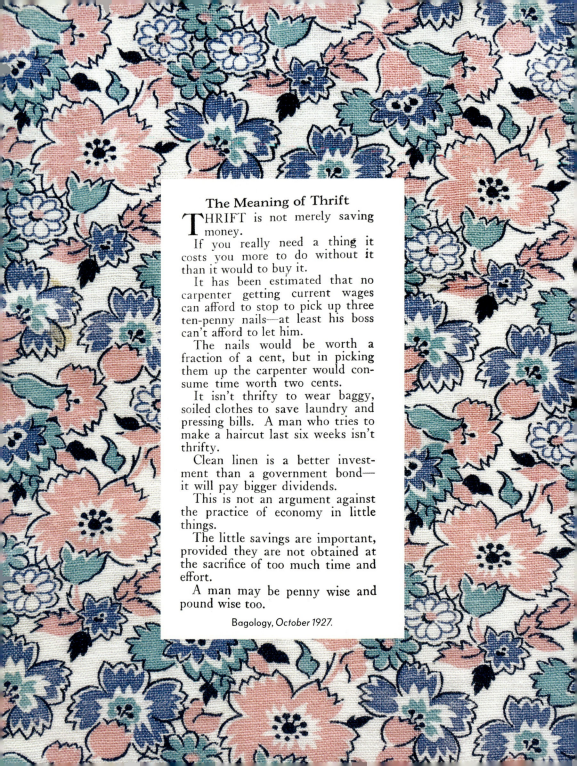

The Meaning of Thrift

THRIFT is not merely saving money.

If you really need a thing it costs you more to do without it than it would to buy it.

It has been estimated that no carpenter getting current wages can afford to stop to pick up three ten-penny nails—at least his boss can't afford to let him.

The nails would be worth a fraction of a cent, but in picking them up the carpenter would consume time worth two cents.

It isn't thrifty to wear baggy, soiled clothes to save laundry and pressing bills. A man who tries to make a haircut last six weeks isn't thrifty.

Clean linen is a better investment than a government bond—it will pay bigger dividends.

This is not an argument against the practice of economy in little things.

The little savings are important, provided they are not obtained at the sacrifice of too much time and effort.

A man may be penny wise and pound wise too.

Bagology, October 1927.

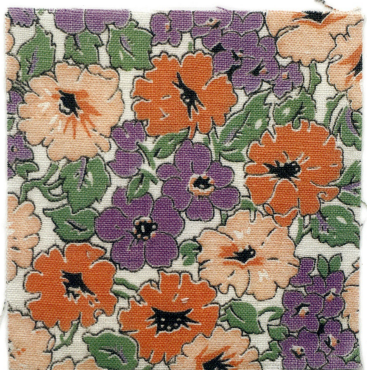

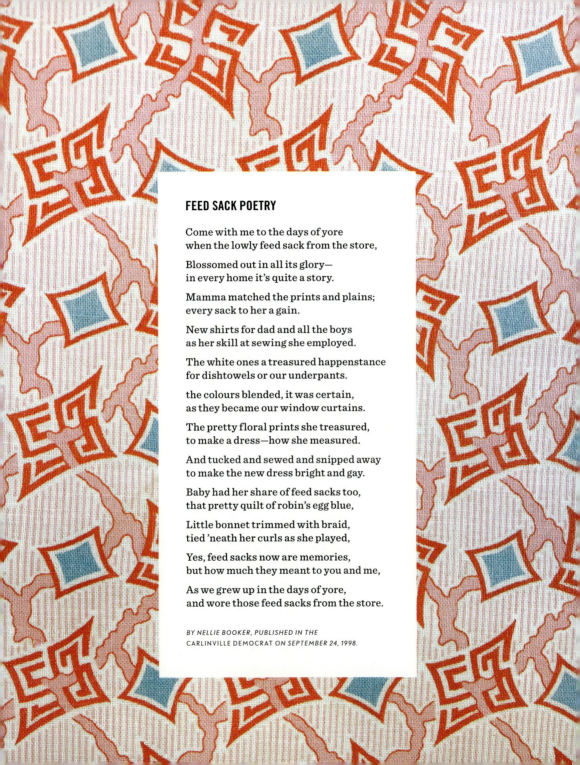

FEED SACK POETRY

Come with me to the days of yore
when the lowly feed sack from the store,

Blossomed out in all its glory—
in every home it's quite a story.

Mamma matched the prints and plains;
every sack to her a gain.

New shirts for dad and all the boys
as her skill at sewing she employed.

The white ones a treasured happenstance
for dishtowels or our underpants.

the colours blended, it was certain,
as they became our window curtains.

The pretty floral prints she treasured,
to make a dress—how she measured.

And tucked and sewed and snipped away
to make the new dress bright and gay.

Baby had her share of feed sacks too,
that pretty quilt of robin's egg blue,

Little bonnet trimmed with braid,
tied 'neath her curls as she played,

Yes, feed sacks now are memories,
but how much they meant to you and me,

As we grew up in the days of yore,
and wore those feed sacks from the store.

BY NELLIE BOOKER, PUBLISHED IN THE
CARLINVILLE DEMOCRAT ON SEPTEMBER 24, 1998.

Flour Prints

She doesn't bake bread,
Rolls, cookies or pie,
Yet she buys SACKS of flour
each day.
Were you to ask, "Why?"
She'd wistfully sigh:
"'Cause the flour sack prints
are so gay!"